PENGUIN BOOKS
CREATION REVISITED

P. W. Atkins is a lecturer in Physical Chemistry at the University of Oxford and a Fellow of Lincoln College, Oxford. He is the author of numerous textbooks, most notably the international bestseller *Physical Chemistry*, first published in 1978 and now in its fifth edition. He is equally well known as an author of outstanding and widely praised science books for the general reader, including *The Second Law* (1984), *Molecules* (1987) and *Atoms, Electrons and Change* (1991). When first published, Dr Atkins's *The Creation* (1981), of which *Creation Revisited* is a revised and extended edition, was hailed as both 'brilliant and challenging' (*The Times Literary Supplement*) and 'exciting and memorable' (*The Times Higher Educational Supplement*). The *Practical Observer* found the new edition 'fabulously thought-provoking'.

P. W. ATKINS

———

CREATION REVISITED

PENGUIN BOOKS

PENGUIN BOOKS

Published by the Penguin Group
Penguin Books Ltd, 27 Wrights Lane, London W8 5TZ, England
Penguin Books USA Inc., 375 Hudson Street, New York, New York 10014, USA
Penguin Books Australia Ltd, Ringwood, Victoria, Australia
Penguin Books Canada Ltd, 10 Alcorn Avenue, Toronto, Ontario, Canada M4V 3B2
Penguin Books (NZ) Ltd, 182–190 Wairau Road, Auckland 10, New Zealand

Penguin Books Ltd, Registered Offices: Harmondsworth, Middlesex, England

First published by W. H. Freeman 1992
Published in Penguin Books 1994
1 3 5 7 9 10 8 6 4 2

Printed in England by Clays Ltd, St Ives plc

For nature is pleased with
simplicity, and affects not the
pomp of superfluous causes.

Isaac Newton
Principia, Book III

PREFACE

My first visit to the Creation took place in 1981. Then I took the view that there is nothing that cannot be understood, and that the path to understanding is to peel away appearances in order to expose the core, which is always of unsurpassed simplicity. I explained that we would travel along a path where we would encounter very simple questions and, more importantly, discover that they have very simple answers. I aimed to show then, as I am to show on this return visit, that it is possible to think rationally about what many regard as lying beyond explanation, such as the processes involved in the creation of the universe and the emergence in it of consciousness.

My aim on the first visit was to argue that the universe can come into existence without intervention, and that there is no *need* to invoke the idea of a Supreme Being in one of its numerous manifestations. I accepted then, as I accept now, that anyone who is in some sense religious is not likely to be swayed by arguments like mine. In this respect I did not intend offence and nor do I intend it now. There were some who took it, though, and I suppose there will be a new generation who will take it again with this revision.

There were two varieties of offence that I could discern and which will probably recur. One was the predictable offence taken by those who considered the great questions I touch on as too tender for human trampling: science for them is a scab and human spirit the tender underside that should be left to lie in peace and not be picked at. I do not budge from the view, though, that the human brain is an instrument of limitless power and that the scientific method (permitting its cautious development), a technique of limitless applicability and (speculatively) limitless success.

The other variety of offence welled in the hearts of scientists.

They considered, either explicitly or inwardly, that my seemingly incautious extension of current scientific thought burst through the membrane of propriety. To them all I can say is that in my words I have sought to excite in my lay readers the sense of power that resides in our noble intellectual pursuit and have sought to show—while emphasizing what is speculative, and freely admitting that almost everything from Chapter 5 on must lie in that category—that the questions that I judge they consider great are *possibly* open to resolution within a paradigm of recognizable science.

There are two principal changes in this revision. I have largely rewritten the left-hand pages of commentary, bringing them up to date and adding further elucidation where it was within my grasp. I have also added a new chapter which explores what I regard as perhaps the deepest of all questions: why mathematics works. It is in this question, I consider, where the twin Amazons of consciousness and physical existence merge, that we shall find the ultimate resolution of the infinitely deep.

In conclusion, I have to admit that I am close to the edge of my own understanding in the speculations that fringe and permeate the later chapters (from Chapter 5 on). I intend these chapters to be only impressionistic passages that convey what I think may prove to be the general form of the resolution of the questions that have driven and plagued humanity since it dropped, bewildered, from the trees. The true, presumably comprehensible, account of these matters will be available only when mathematically austere versions of them become available. That, I regret, is beyond me.

P.W.A.

Lincoln College
Oxford
1992

CONTENTS

One

OBVIOUS THINGS

I SHALL TAKE YOUR MIND ON A JOURNEY. IT IS A JOURNEY OF comprehension, taking us to the edge of space, time, and understanding. On it I shall argue that there is nothing that cannot be understood, that there is nothing that cannot be explained, and that everything is extraordinarily simple.

A great deal of the universe does not need any explanation. Elephants, for instance. Once molecules have learnt to compete and to create other molecules in their own image, elephants, and things resembling elephants, will in due course be found roaming through the countryside. The details of the processes involved in evolution are fascinating, but they are unimportant: competing, replicating molecules with time on their hands will inevitably evolve.

Some of the things resembling elephants will be men. They are equally unimportant. It is undeniable (but not necessarily predictable) that molecules, once they have stumbled upon reproduction, will, somewhere or other (here, as it happens), band together into corporations shaped into the form and having the functions of men, and that these men will also one day be found roaming through some countryside. Their special but not significant function is that they are able to act as commentators on the nature, content, structure, and source of the universe and that, as a sideline, they can devise and take pleasure from communicable fantasies.

Molecules equipped for competition, survival, and reproduction are also devoid of significance. It is undeniable that they can emerge given the right mixture of ingredients, a stable warm platform, and time.

Small molecules evolve into bigger ones by eating smaller ones, although it is not always clear which is the eater and which the eaten. Little molecules eat by impact, and what emerges from their collisions is sometimes a bigger molecule—more atoms strung together—or an atom in the original molecule replaced by a cluster of atoms in the new. Sometimes an almost complete molecule

I shall say much more about chemical reactions in, due course. Reaction in solution is much more subtle than in a gas, just as gardening gives more subtle products than traffic accidents. In solution there is the possibility of complexity on account of the subtlety that is open to a reaction—such as the modification of one or two bonds in a molecule built from thousands.[5] For an introduction to the concepts involved in the discussion of chemical reactions, see my *Atoms, electrons, and change*.[6]

An introductory account of the formation of the elements—nucleosynthesis—can be found in *The Cambridge Encyclopaedia of Astronomy*.[7] In brief, the hydrogen and helium in the universe were synthesized in the initial explosion, the 'big bang'. Heavier species have been (and still continue to be) formed by roasting in the interiors of stars and then distributed over the rest of the universe in the explosions that occur at various stages of a star's life cycle. Another very readable account is the book by Cox.[8]

remains incorporated into the network of atoms of the original molecule, like a fly in a web. The winners of such dinners can in turn go on to others, and successful diners develop more sophisticated ways of trapping dinners. In due course the sophistication becomes so great that the really successful keep the less in herds, and devouring is governed by philosophical and economic speculation.

Even little unsophisticated primitive molecules are insignificant. It is undeniable that they can emerge into existence if the appropriate atoms are available, for little molecules are just a few atoms stuck together, and atoms stick together. If there are atoms there will in due course be molecules; and if there are molecules on warm, wet platforms, there will in due course be elephants.

I am sure you can see where we are heading. Suppose you were to set about designing a universe. If you were omnipotent you could draw up a detailed specification of all creatures great and small. If you were designing our present universe you would include a specification for an elephant. But elephants have turned out to be inevitable given molecules capable of competition and reproduction and experiencing the environmental history of this planet. Therefore, if you were not pressed for time, a much simpler approach would be to specify an array of fighting molecules, bring them together, and sit back and wait. After a while, their descendants would be elephants, and men.

Yet complex molecules emerge from simpler ones inhabiting planets; and so the specification could be simplified still further. That simplification can itself be simplified, for if you merely specify the elements, and perhaps some other things, sooner or later there will be elephants.

The question that now arises is the following. Suppose you prefer to be an infinitely lazy creator: what is the *minimum* specification you can get away with? Need you really go to the trouble of specifying a hundred or so different kinds of atom? Is it possible to specify a mere handful of things, which, if they exist in the appropriate amounts, lead first to elements and then to elephants? Can the whole of the universe be taken back to a *single* thing, which, if it is appropriately specified, leads inevitably to

No one is yet in a position to be definite about the final solution of the problem of cosmogony (the generation of the universe), and so I have to warn the reader that the argument will increasingly resort to speculation. While it is quite easy to use straightforward language to express ideas that are already in some sense established it is difficult to simplify expositions of ideas that have not yet been formulated exactly.

There are two consequences for the present book. One is that speculation and fact will be tangled together. I shall try to indicate which is which. Another is that there will almost certainly be a sense of disappointment at the end (at least). This is because we do not yet know the whole truth about cosmogony, and therefore the account can hardly be expected to be other than lacking in detail. What I am trying to do, it should always be remembered, is to show that deep questions about cosmogony can be asked, that in some cases they have already been answered, and that in others science is pointing to the type of answer that may be forthcoming quite soon.

An introduction to the composition of the universe, and the way it is determined is given by Kaufmann.[9] A general, popular, and very engaging view of the composition of the universe, including a simple account of the origin of the elements, is given by Nigel Calder.[10] My allusion to galactic dust is probably a memory of his more evocative 'In a sense human flesh is made of stardust' (p. 32).

elephants? Could you (being infinitely lazy) avoid, in fact, specifying and making even that? If you could (and we shall come close to seeing that you can), there would be no role for you in the creation of your universe.

Our task should by now be clear. We have to embark upon the track of the absolute zero of creative involvement in the creation, the absolute zero of intervention. The only clue we have at the outset is that the final answer will almost certainly be one of extreme simplicity, for only the perfectly simple can come into existence while all agents sleep (or are not there). This suggests that we should examine the universe for the footprints of its underlying simplicity. As we look for them we must always remember that complexity of behaviour and appearance may be illusions, and what we perceive as complexity may be the outcome of chains of simplicity.

This is where we begin. The only faith we need for the journey is the belief that everything *can* be understood and, ultimately, that there is nothing to explain.

Our appreciation of the nature of the universe stems from our ability to notice, observe, and reflect on the things it contains. We notice, for instance, that everything is built from the same stuff. Animals eat plants and drink rivers. Plants eat mountains. When animals die they contribute to later mountains and other plants. Mountains sprout from planets which are the accretions of debris from dead stars. Everything is built from the same stuff, and the further we look into the distance the less likely it appears that any different stuff is involved elsewhere. We are galactic dust, and to galactic dust we shall return.

We notice there is a universe. I mean by that much more than that there is a collection of stars floating in a void and containing us. One of the greatest of discoveries has been that the universe is open to measurement, and that there is meaning in the contemplation of its range and its age. There is more of a revolution in the realization that the universe can be measured than in its actual measurement, for the existence of duration confronts us with the problems of a boundary to time. Noting this problem is a step towards understanding, because if we can comprehend what it

For an assessment of the size of the universe, how it is measured, and a catalogue of its contents, the *Cambridge Encyclopaedia* is as good a source as any. A recipe for one model universe is available[11] on a scale undreamt of by Mrs Beeton. On the basis that the present age is 10^{10} years, the present radius is 1.3×10^{10} light-years (1.2×10^{26} m). The mean density now is about 1.4×10^{29} g cm^{-3} (corresponding to about one atom in a cubic metre on average), and its total mass is about 5.7×10^{56} g. There are about 29×10^{23} stars distributed as about 10^{11} galaxies. Every 5 seconds the universe expands by an amount approximately equal to the volume of our galaxy. It will all be over in about 5×10^{10} years or so. Critical analyses of these quantities are also available.[12, 13]

This is Olbers' Paradox.[14, 15] The paradox arises on the basis that stars have finite sizes, and any straight line drawn from an observer and projecting into infinite space will sooner or later intersect the surface of the star. This presumes that the universe is infinite in extent, infinitely old, uniform in space, and static. The paradox could be resolved in a variety of ways. For instance, if the universe is finite, then there will be no stars beyond a particular radius, and hence some lines from an observer's eye will not touch a star. Likewise, if the universe is infinite but of finite age, the light from distant stars would not yet have had time to reach us. The modern resolution of the paradox is more subtle, like so much in general relativity and cosmology where familiar concepts like age and distance lose their simplicity. The modern resolution is based on discarding the assumption that the universe is static. For an entrance into the debate about the validity and resolution of the paradox, see reference 16.

means to be before the start of time, then we shall be closer to comprehending the nature of time itself. The key to understanding lies in the identification and comprehension of the most primitive.

Whether or not we regard the discernible universe as big or small is immaterial. On the scale of man's size it is certainly vast. But man is not intrinsically significant, and so we cannot regard him as a significant criterion of size. The vastness of the universe is tamed if we have confidence to think about it on a large-enough scale: a big-enough attitude of mind dispels the awe that vastness inspires. Awe stultifies. Think of the universe as a puff of dust about a metre in diameter. Every dust grain is a galaxy. We live near a rather ordinary star which is a member of a rather ordinary galaxy somewhere insignificant in the puff of dust.

Every night we are shown that the universe had a beginning, but most of us simply regret, use, or enjoy the dark without perceiving that it brings knowledge.

A moment's thought is enough to show that the darkness of night eliminates one half of eternity. If the universe were infinite and eternal, then in whichever direction we looked we would see the light of a star. Every point of the sky would be a star, and the entire sky would be as brilliant as the surface of the sun. Even by day the sun would be indistinguishable from the brilliance of its background. But, at night, the sky is dark, and there is an in-between the stars; so the universe cannot be both infinite and eternal. Here is an example of how noticing a commonplace could have inspired a revolution.

A sharper intelligence could cut more knowledge from the dark. It could perceive how an overall expansion of the universe can stretch the light of distant stars and diminish their intensity. As it was, astronomers saw this expansion through their telescopes before the sharper intelligence cut. They saw the galaxies receding, the cloud of dust expanding. Nothing is more natural than to pursue the expansion back into the past, and to imagine the cloud of dust emerging from an explosion at its core. That explosion was creation.

But what was it that was created? What is it that is expanding? Is it the galaxies spreading out through space? But, then, what is

The electron microscope, when used in an adapted form, is now sensitive enough to give images of atoms. X-ray diffraction is the basis of the modern technique of determining the structure of molecules, especially the huge molecules of biological importance such as proteins and DNA. It shows electrons grouped into blobs clearly identifiable as atoms. Field ionization microscopy[17] gives representations of individual atoms, and in an adaptation can be used so precisely that the experimenter can locate a single atom, pluck it from the surface of the sample, and identify it or use it in some way. Some of the most astounding images of atoms and molecules have come from the recent development of scanning tunnelling microscopy, in which a fine point—a point so fine, indeed, that it ends in a single atom—is moved in parallel lines across the face of a sample, and the atoms are revealed by the current that flows through the tip.[17] An adaptation of this technique can be used to move single atoms into arbitrary but intentional locations on the surface.

The diameter of an atom is of the order of 2×10^{-10} m. This size is governed by the strength of the electrostatic interaction between the nucleus and the surrounding electrons. The strength is governed by a fundamental constant α. Its value is about 1/137. The diameter of an atom is inversely proportional to α, and so if it was twice its present value, we would be only half our present size, eight times denser, and more vivid.[18]

space? What does it occupy? Where did it come from? Is space itself expanding? Was it merely space that was created? What is space expanding into?

We notice, we think, that the universe is not only space. We, at least, are in it, and there is other matter. A comprehension of the nature of the world must find a place in it for atoms and (in the general sense) for elephants, and for substance and (in a guarded sense) for spirit. Somehow matter must have been created out of something resembling nothing. Happily, though, our demands on the complexity of the original creation have diminished as scientific investigation has peeled back the layers of the onion we refer to as the structure of matter. No longer do we have to account for the generation of elephants, with all their inner complexity. Now the problem is the generation of the components of atoms.

We know that atoms exist because we can see them. Sophisticated versions of microscopes give us images of atoms, and let us photograph molecules. We can crack open atoms and look inside.

Atoms are very big: they have to be, because so much is packed away inside. It is true that we often think of them as very small; but that is only because we are very big (for much the same reason). Just how big an atom is can be imagined by thinking of it as expanding until the size of the nucleus at its core is the size of a man, then the atom itself would be a thin haze of electrons spreading through a 100-km sphere around it.

The diffuseness of the outer parts of atoms shows the weakness of the control of the central nucleus over the electrons that surround it. That weakness underlies the richness of life. It means that atoms can be dislodged from molecules with only gentle persuasion, and that new arrangements of atoms can unfold out of old. On account of the weakness, structures are not frozen into unchangeable arrays, but can respond to their surroundings. There is responsiveness in the loosely bound structures of atoms and molecules, and change can occur when the environment gently prods. If structures had been tighter, prods like nuclear explosions would have been the only way of achieving change, and none of the subtlety of perception and consciousness could have emerged. Evolution would have been as destructive as it is constructive.

The numbers of atoms required in order to result in a complex creature is discussed in general terms and somewhat simplistically in a classic little book by Schrödinger.[19]

The rate of evolution has been summarized by Freeman Dyson as follows: 'Looking at the past history of life, we see that it takes about 10^6 years to evolve a new species, 10^7 years to evolve a genus, 10^8 years to evolve a class, 10^9 years to evolve a phylum, and less than 10^{10} years to evolve all the way from primaeval slime to Homo Sapiens.'[20]

Other things being equal (and in particular the temperature of the environment being independent of α), if α had been 1 per cent larger than it is, then man would have taken twice as long to evolve. If α had been twice as big as it is, then instead of evolving in 10^{10} years man would have needed 10^{62} years, which is far longer than the present age of the universe (10^{10} years). The forces acting within nuclei are about 100 times stronger than the electrostatic force, and negligible amounts of change would have occurred except in regions where the temperature is extremely high, such as at the centres of nuclear explosions and inside stars. There are numerous consequences of changing the value of α, and we shall encounter more as we go on.[21, 22]

See J. C. Polkinghorne's book[23] for a down-to-earth survey of modern particle physics. The author wrote it in the course of his transition from a chair of mathematical physics to membership of the Anglican priesthood. An easier introduction, written with great style and imagination, is the book by Nigel Calder[10] which we mentioned above. It is particularly good on the nature of forces and the role of ever more elementary particles. For a history of forces, see the enthralling volume by Pais.[24]

At the same time as the frailty of molecular structure lets matter respond to the gentle demands of its environment and evolve subtlety by acquiring complexity, it continues to contribute to the cultural complexity of the species. So long as the molecules in the organism can respond to the influences of its surroundings, the organism can observe. The consequences of observations are the perceptions and inventions of the mind, which in turn are manifestations of shifts of atoms and modulations of molecular structure within the brain.

The sensitivity of response has as its disadvantage transience. Small departures from restraint are enough to kill. Warmth that warmed can, beyond a threshold, scorch and burn. That is why it is so easy to die.

Whereas the outer parts of atoms are regions marshalled by gentle forces, the existence of a nucleus shows that there are stronger forces operating in the atom's core. Only a much stronger force can grip the components of the nucleus into a dense bundle. That is why nuclear change is both energetically and economically more vigorous than chemical, and why the timid prodding of the alchemists failed to transmute the elements.

The strong force that binds the nucleus must have a range that is very short, for otherwise the whole universe would already have been wound down into a single blob. Moreover, while the force has mastery over the little nuclei of light atoms, at uranium, with a couple of hundred particles packed into the nucleus, it is losing control, and the nucleus tends to fall apart. Whether it does so in a controlled or an uncontrolled way determines the level of social benevolence of the outcome.

Elephants have a deeper structure, their corporation of molecules. Molecules have a deeper structure, their atoms weakly stuck together. Atoms have a deeper structure, a cloud of electrons and at the heart a nucleus. Nuclei have a deeper structure, the cluster of protons and neutrons strongly bound together. Is the onion infinite?

It appears not. One more layer is known, and is found when you try to crack open protons. You find you cannot (which is significant), but you can discern inside them their constituents the

There must be some kind of inner structure so as to differentiate between the many kinds of quarks that have been identified. One of the things we must do in what follows is identify some kind of primitive component. A view on this will emerge in the following pages. These remarks do not rule out the possibility that the truly ultimate entities are organized into groupings of a kind more primitive than their grouping into quarks, and therefore that the quarks themselves may have some kind of internal structure even though they lack spatial extent. Such inner entities have been termed 'preons'.[25]

There are two categories of name. One includes terms that gives us the impression of knowing what we mean: I have in mind 'familiar' concepts such as electric charge. Everyone knows what is meant by electric charge (until they are asked to explain it). The other category includes terms which we immediately accept as unfamiliar: by an ironic twist they are given homely, familiar names like 'colour', 'flavour', 'strangeness', and 'charm'. Some people object to the jocularity of this nomenclature, but it is good in two senses. One is that it is fun, which is not out of place in science. The other is that such names signal unambiguously the fact that they are code words needing fuller elucidation

quarks. Like electrons, the quarks appear to be without spatial extent and to lack a deeper structure: they have character without extension and are substance without inside. Here we must be on the brink of ultimate simplicity, almost at the end of the onion, for anything with structure is too complicated to be regarded as surpassingly simple. Complexity, even of the most rudimentary kind (such as extent in space), must be an attribute of combined simplicities, and not something that can emerge fully made up.

I think we ought to pause here and assess our position. The recipe for the universe is on the way to being simple. We appear to need quarks, electrons, maybe a few other things, and a variety of forces to hold everything together with various degrees of rigidity. Nevertheless, the recipe is still both overcomplicated and insufficient. It is overcomplicated because it demands the specification and creation of at least half a dozen attributes of the universe, which is not what is meant by exceedingly simple. Moreover, even its apparent simplicity is fraudulent because it hides complications behind names. Only when we know what words like 'force', 'electron', and so on truly represent can we claim true comprehension. Names are codes; we should not let familiarity with them masquerade as understanding.

The way ahead is clear. We must examine the meaning of the simple, familiar concepts that have crept into the account and are normally taken for granted. I shall argue that most of the words are empty of meaning at the deepest level of comprehension, and that they have acquired apparent meaning in order to encapsulate phenomena and facilitate conversations. I shall try to show that most of the concepts that at first sight seem crucial to our understanding of the workings of the world can be allowed to evaporate. When they have gone they will leave behind what we should truly and usefully consider: that is, virtually nothing.

First Orientation

We are on a journey to discover both the ultimate nature of the universe and the manner in which it came into being. I am taking the view that the ultimate fabric must be of extreme simplicity, and that its perceived complexity and richness must be a result of primitive things grouping together into gangs. I am also taking the view that only the very simplest things could come into existence at the creation, and therefore that the job of any creator was slight. I am developing the opinion that the only way of explaining the creation is to show that the creator had absolutely no job at all to do, and so might as well not have existed. We can track down the infinitely lazy creator, the creator totally free of any labour of creation, by resolving apparent complexities into simplicities, and I hope to find a way of expressing, at the end of the journey, how a non-existent creator can be allowed to evaporate into nothing and to disappear from the scene.

In order to draw attention to the insights provided by the commonplace, and to show how apparent complexity is ultimately organized simplicity, we now turn to the causes of change. I shall argue that all forms of change, from the rudimentary, such as cooling, to the exceedingly complex, such as the formation of opinions, are eruptions into the perceived world of the same primitive events. We shall see that all the events around and inside us have the same motivation: they are driven by a purposeless collapse into chaos.

Two

WHY THINGS CHANGE

Quality? What is quality? This is a term introduced by Freeman Dyson.[27] For the time being, regard localized energy as energy with potency to be harnessed to do work, and therefore in some sense having 'high quality'. Work involves ordered motion; heat involves random motion. In the process of any change, highly localized energy becomes more dispersed, and we are no longer able to point to a well-defined location. Dyson (on his p. 52) establishes an 'order of merit' for various forms of energy. Gravitational energy heads the list and has the highest quality. At the bottom stands the cosmic microwave radiation. The latter corresponds to the ultimate heat sink, and there appears to be no way in which this energy can be further degraded.

CHANGE TAKES A VARIETY OF FORMS. THERE IS SIMPLE CHANGE, as when a bouncing ball comes to rest, or when ice melts. There is more complex change, as in digestion, growth, reproduction, and death. There is also what appears to be excessively subtle change, as in the formation of opinions and the creation and rejection of ideas. Though diverse in its manifestations, change does in fact have a common source. Like everything fundamental, that source is perfectly simple.

Organized change, the contriving of some end, such as a pot, a crop, or an opinion, is powered by the same events that stop balls bouncing and melt ice. All change, I shall argue, arises from an underlying collapse into chaos. We shall see that what may appear to us to be motive and purpose is in fact ultimately motiveless, purposeless decay. Aspirations, and their achievement, feed on decay.

The deep structure of change is decay. What decays is not the quantity but the *quality* of energy. I shall explain what is meant by high quality energy, but for the present think of it as energy that is localized, and potent to effect change. In the course of causing change it spreads, becomes chaotically distributed like a fallen house of cards, and loses its initial potency. Energy's quality, but not its quantity, decays as it spreads in chaos.

Harnessing the decay results not only in civilizations but in all the events in the world and the universe beyond. It accounts for all discernible change, both animate and inanimate. The quality of energy is like a slowly unwinding spring. The quality spontaneously declines and the spring of the universe unwinds. The quality spontaneously degrades, and the spontaneity of the degradation drives the interdependent processes webbed around and within us, as through the interlocked gear wheels of a sophisticated machine. Such is the complexity of the interlocking that here and there chaos may temporarily recede and quality flare up, as when cathedrals are built and symphonies are performed. But these are temporary and local deceits, for deeper in the world the

The idea pervading this chapter is *entropy*.[17, 26] This is the world of the Second Law of thermodynamics, and is interpreted in any survey of the molecular basis of thermodynamics. The Second Law of thermodynamics, a summary of experience about the direction of change, is consistent with the remark that the entropy of the world increases whenever there is a change. This law in turn is consistent with the remark that the number of ways of distributing the available energy increases.

That random jostling leads to irreversible change can be understood on the basis of a random walk. A simple version is as follows. Consider an object that is free to step with equal probability to the left or to the right. Although we may not be in a position to observe the direction taken by an individual particle on each step we can at least predict the probability that some particle will be found at some distance from its origin after it has had time to take a specified number of steps. If we consider a crowd of objects initially all heaped together at some location, then since they are all continuously leaping at random to the left or right, some will be found at a great distance at a later time (for instance it is possible, but improbable, that one object took every one of its steps to the right), while most will be found close to the origin. The peak of population at the origin continues to spread with time, and there is negligible probability (but actual possibility) that all the members of the original crowd will simultaneously be found back in their original heap. Hence we have an irreversible change by virtue of random leaping. Elaborations of this simple idea can be explored into various intricacies.[28, 29]

spring inescapably unwinds. Everything is driven by decay. Everything is driven by motiveless, purposeless decay.

As we have said, by 'quality' of energy is meant the extent of its dispersal. High quality, useful energy, is localized energy. Low quality, wasted energy, is chaotically diffuse energy. Things can get done when energy is localized; but energy loses its potency to motivate change when it has become dispersed. The degradation of quality is chaotic dispersal.

I shall now argue that such dispersal is ultimately natural, motiveless, and purposeless. It occurs naturally and spontaneously, and when it occurs it causes change. When it is precipitate it destroys. When it is geared through chains of events it can produce civilizations.

The naturalness of the tendency of energy to spread can be appreciated by thinking of a crowd of atoms jostling. Localized energy, energy in a circumscribed region, corresponds to vigorous motion in a corner of the crowd. As the atoms jostle, they hand on their energy and induce their neighbours to jostle too, and soon the jostling disperses like the order of a shuffled pack. There is very little chance that the original corner of the crowd will ever again be found jostled back into its original activity with all the rest at rest. Random, motiveless jostling has resulted in irreversible change.

This natural tendency to disperse accounts for simple processes like the cooling of hot metal. The energy of the block, an energy captured in the vigorous vibrations of its atoms, is jostled into its surroundings. The individual jostlings may result in energy being passed in either direction; but there are so many more atoms in the world outside than in the block itself, that it is much more probable that at all later times the energy of the block will be found (or lost) dispersed.

Illusions of purpose are captured by the model. We may think that there are reasons why one change occurs and not another. We may think that there are reasons for specific changes in the location of energy (such as a change of structure, as in the opening of a flower); but at root, all there is, is degradation by dispersal.

Suppose that in some region there are many more places for energy to accumulate than elsewhere. Then jostling and random

23

Energy can be regarded as coming in bundles, 'quanta', and is stored in the various types of motion of molecules and the way their atoms are arranged.[18] Whether or not a molecule, or a collection of molecules, can store a lot of energy depends on its constitution. A complex molecule can be regarded as a labyrinth for energy: there are so many ways that energy can be located that it can be thought of as jostling around inside the molecule for a long time before escaping into the surroundings. Therefore if energy jostles from the surroundings at random, it will appear to be trapped inside the molecule. The energy is nevertheless dispersed, and so large is the number of places that it can be located within the molecule that it gives the impression of favouring location there at the expense of elsewhere.

All natural change corresponds to the increase of entropy, and the underlying basis of the apparent irreversibility of the history of the universe[29, 30] is the extreme improbability that energy, atoms, and molecules will attain earlier locations and configurations.

leaping results in its heaping there. If the energy began in a heap initially organized elsewhere, it will be found later in a heap in the region where the platforms are most dense. A casual observer will wonder why the energy chose to go there, conclude there must have been a purpose, and try to find it. We, however, can see that achieving being there should not be confused with choosing to go there.

Changes of location, of state, of composition, and of opinion are all at root dispersal. But if that dispersal spreads energy into regions where it can be located densely, it gives the illusion of specific change rather than mere spreading. At the deepest level, purpose vanishes and is replaced by the consequences of having the opportunity to explore at random, discovering dense locations, and lingering there until new opportunities for exploration arise.

Events are the manifestations of overriding probabilities. All the events of nature, from the bouncing of balls to the conceiving of gods, are aspects and elaborations of this simple idea. But we should not let pass without emphasis the word probability. The energy just might by chance jostle back into its original heap, and a structure reform. The energy just might, by chance, jostle its way back into the block from the world at large, and an observer see a cool block spontaneously becoming hot or a house of cards reforming. These possibilities are such remote chances that we dismiss them as wholly improbable. Yet, while improbable, they are not impossible.

The ultimate simplicity underlying the tendency to change is more effectively shrouded in some processes than in others. While cooling is easy to explain as natural, jostling dispersal, the processes of evolution, free will, political ambition, and warfare have their intrinsic simplicity buried more deeply. Nevertheless, even though it may be concealed, the spring of all creation is decay, and every action is a more or less distant consequence of the natural tendency to corruption.

The tendency of energy to chaos is transformed into love or war through the agency of chemical reactions. All actions are chains of reactions. From thinking to doing, in simply thinking, or in responding, the mechanism in action is chemical reaction.

Why things change

As an example of a straightforward chemical reaction, take the combustion of a lump of coal. In its simplest aspect the reaction is the combination of the carbon atoms in the coal with the oxygen molecules of air to form carbon dioxide (CO_2). The reaction releases a considerable amount of energy. When a pair of oxygen atoms and a carbon atom get into an arrangement roughly corresponding to a CO_2 molecule, their newly released energy is in the form of vigorous vibration. This vibrational energy is dissipated very rapidly as the atoms in the lump are knocked into vibration themselves, and the energy spreads into the lump and away from the site of reaction. The carbon and oxygen atoms are now trapped into being a CO_2 molecule. The reason why carbon dioxide does not spontaneously settle out into a layer of soot and a cloud of oxygen is that although the reverse of the process just described *may* occur, it will do so with only vanishingly small probability. This is because, in order to form soot from carbon dioxide, energy has to be localized in the molecule from its surroundings; but it is very unlikely that enough energy will spontaneously and simultaneously accumulate in the tiny CO_2 molecule.[26]

At its most rudimentary, a chemical reaction is a rearrangement of atoms. Atoms in one arrangement constitute one species of molecule, and atoms in another, perhaps with additions or deletions, constitute another. In some reactions a molecule merely changes its shape; in some, a molecule adopts the atoms provided by another, incorporates them, and attains a more complex structure. In others, a complex molecule is itself eaten, either wholly or in part, and becomes the source of atoms for another.

Molecules have no inclination to react, and none to remain unreacted. There is, of course, no such thing as motive and purpose at this level of behaviour. Why, then, do reactions occur? At this level too, therefore, there can be no motive or purpose in love or war. Why then do they occur?

A reaction tends to occur if in the process energy is degraded into a more dispersed, more chaotic form. Every arrangement of atoms, every molecule, is constantly subject to the tendency to lose energy as jostling carries it away to the surroundings. If a cluster of atoms happens by chance to wander into an arrangement that corresponds to a new molecule, that transient arrangement may suddenly be frozen into permanence as the energy released leaps away. Chemical reactions are transformations by misadventure.

Atoms are only loosely structured into molecules, and explorations of rearrangements resulting in reactions are commonplace. That is one reason why consciousness has already emerged from the inanimate matter of the original creation. If atoms had been as strongly bound as nuclei, the initial primitive form of matter would have been locked into permanence, and the universe would have died before it awoke.

The frailty of molecules, though, raises questions. Why has the universe not already collapsed into unreactive slime? If molecules were free to react each time they touched a neighbour, the potential of the world for change would have been realized long ago. Events would have taken place so haphazardly and rapidly that the rich attributes of the world, like life and its own self-awareness, would not have had time to grow.

The emergence of consciousness, like the unfolding of a leaf, relies upon restraint. Richness, the richness of the perceived world

In saying that energy holds the key to its own degeneration, I have in mind the *activation energy* of reactions.[6] A chemical reaction may take place if two conditions are fulfilled. The first is that the reactant species meet; the second is that when they do so they possess enough energy, the 'activation energy', to undergo reaction. The probability that they actually possess at least that amount of energy is expressed by a formula known as the *Boltzmann distribution*. This formula is a probabilistic result derived on the assumption that energy is distributed randomly among all available modes of motion. The reason why the activation energy stabilizes form is that the Boltzmann expression gives a very small value for the probability that molecules have enough energy to react at normal temperature.

Another form of constraint on the degradation of energy is described by Dyson in the article referred to earlier.[27] He describes a variety of 'hang-ups', or obstacles to degradation that arrest its precipitate collapse. One of these is the 'size hang-up', which is the intrinsic slowness of the collapse of a diffuse object being pulled together gravitationally. Since the universe is so diffuse (about one atom per m^3 on average), it takes a long time to collapse (about 10^{11} years). Another obstacle is the spin hang-up, which arises from the slowness with which rotational motion can be dissipated: among other things it is responsible for the existence of planets in orbits. The third obstacle is the 'thermonuclear hang-up', which prevents stars collapsing beyond a certain point until all their hydrogen has been burnt. This hang-up has given the sun a life of 4.5×10^9 years so far, with an expectation of a further 5×10^9 to come. Fourthly, there is the 'weak-interaction hang-up', which prevents the sun (and any star) from simply going off like a bomb. Dyson continues the list of hang-ups: the article is most engaging (and there is still time to read it).

and the richness of the imagined worlds of literature and art—the human spirit—is the consequence of controlled, not precipitate, collapse.

Energy itself holds the key to its own degradation. Molecules have the opportunity to react when they meet, but they actually do so only if their atoms are loose enough to wander into new arrangements and to expose themselves to opportunities for misadventure. While frail, molecules are not floppy.

In order to explore, the atoms of molecules must be marginally loosened. They are loosened if energy enters the molecule and stimulates vibration, because a vigorously vibrating molecule is a loose cluster of atoms. And how does the energy enter the molecule? It enters by chance. By chance energy may jostle its way in and be there at the moment the molecules happen to meet. By chance a pair of molecules may meet while they happen to be favoured with more than the average share of energy. Then their atoms may wander, and by wandering, react.

I should like to pause at this point to summarize the argument so far. We have seen, at a molecular level at least, how chaos both drives and restrains the world. Collapse into chaos motivates change, for all natural events are outcomes of the tendency to dispersal. Chaos also stabilizes form, because the chance is only small that molecules are favoured with enough energy for them to explore possible alternative arrangements. We are both led on and held back by chaos: chaos is both the carrot and the cart.

If everything, both structure and change, is the outcome of chance orchestrations of chaos, there must be chains linking the superficial to the deep. I should like to try to indicate their nature, if not their detailed form.

Evolution is reaction by seduction. Complex molecules can acquire even greater complexity in stages instead of attempting a single grand passion. One molecule may be able to discard a few atoms to a congenial partner, pick up a few others elsewhere, and in due course chance upon a destination. Only a little reorganization has to take place at each step, and so only a little loosening is required for each one. Since small chance influxes and abundances of energy are more likely to occur than big ones, the overall process

Although one chemical reaction (or a more straightforward physical process) may result in a decrease of the entropy of the universe, and therefore corresponds to a retraction of the dispersion of energy, it may still be driven forward if it is linked in some way to another reaction that does have a tendency to occur naturally, and which can lead *overall* to a net dispersal of energy. The construction and sustenance of complex organisms is due to the constructive reactions being driven by the reactions made available by the ingestion of food.[26]

The molecular basis of replication (in essence the chemistry of DNA) is described in a wide variety of places, such has been the remarkable impact on science and the general public, of the discovery of the molecular basis of life.[5, 31, 32]

A point to bear in mind about the complexity of construction of the brain[33, 34] is that a human brain is thought (by itself; we shall see more of self-reference later) to consist of around 10^{11} neurons. A typical neuron has anything from 1000 to 10 000 synapses (points of contact) with other neurons.

may occur much more quickly than if enough energy had to arrive for there to be reaction in a single stride. That is reaction by multiple misadventure, reaction down the slippery slope. Whether or not the reaction can proceed then becomes mainly a matter of logistics, or the supply of little molecules at the appropriate time in the meal.

The whole course of evolution can be regarded as a geared and cooperative dissipation of energy. Every stage of evolution, including the steps that gave rise to complex molecules out of simpler ones, to people out of slime, and the processes involved when species are confronted with competition, proceeds by dissipation.

Molecules did not aim at reproduction: they stumbled upon it. Accretion of complexity reached the point where one molecule was so structured that the sequence of reactions it could undergo, under the casual pressure of dispersal, led by chance to the formation of a replica. That molecule naturally had the same reproductive ability, and even though the first might have been eliminated by the evaporation of a pond, the daughter could continue the line. At every stage of replication there was opportunity for modification because slightly different smaller molecules were in the vicinity and could be incorporated. Many of these daughters may have been unviable, or less successful at replication than their ancestors and sisters; but some were more successful, and flourished into elephants.

Perceptions of the external world developed in subtlety with the evolution of the complexity of the body. Those perceptions, like decisions to embark upon activities and our reflections on our own activities and those of others, are all manifestations of reactions. We interact with the external world as the breeze of events shifts specially responsive groups of atoms fashioned into eyes and ears.

Since reactions are aspects of chaos, perceptions, decisions, and reflections are also ultimately driven by an underlying tendency to chaos. The apparent complexity of consciousness is the outcome of the complexity of the interdependence of the reactions geared to this decay, and there is no need to regard it as an intrinsic complexity embellished by a soul. Behaviour is the complex organization of simple processes, and the complex structure of the brain is the complex gearing that marshals simplicity into apparent complexity.

The retina consists of about 3×10^6 cones (which are responsible for colour vision) and 10^9 rods (which do not distinguish colour). The remarkable refinement of the eye (which has evolved independently about half a dozen times—on this planet alone) is reflected in the fact that the rods can respond to a single photon of light. The two significant features of the antenna molecule in the rods is that it is bent in the middle and that it absorbs light of wavelengths similar to the sun's (which is why we can see in daylight). The primary event in vision is that this molecule absorbs a photon (if one is focused on it), the bend unbends and uncoils into a more or less straight stick.[5] This stick does not fit into the available space. The change of shape affects the ability of sodium and potassium ions to cross the nerve membrane attached to one end of the rod. We shall see in a moment that the balance of concentrations of these ions is central to the propagation of nerve impulses.

The axons of nerve cells (along which the impulse runs) are bathed in a fluid which is rich in sodium ions; the inner fluid is rich in potassium.[33] On account of this imbalance of concentrations there is a potential difference between the inside and the outside of the nerve cell. The nerve impulse arises as channels in the membrane open like lock gates and allow sodium ions to flood in. Then the channels close, and another set open. These allow the potassium ions to flow out, with the result that the original voltage is restored, but at the expense of losing the original sodium and potassium concentration differences. The spike of changing voltage ripples along the nerve and enters the synapse. There molecules of a chemical transmitter, stored in bags, are released into the next neuron. These chemicals can either enhance the neuron's likelihood of firing an impulse through its connections to other cells, or they can inhibit it. This is the complex interplay of switching functions within the brain; everything is chemical, and everything moves in the direction of increasing entropy.

The structure ensures that simple chemical processes within the cells of the brain are coordinated into a whole which is both sufficiently complicated to be rich in properties and sufficiently unpredictable to encompass imagination and invention.

Take perceiving. Its essence is the acquisition of information about events external to the brain's bearer, and events not wholly external, as in pain. Bodies have antennae—nerve endings—that respond to their environment and capture information. These sensors, bunched into things like eyes, trigger signals to the brain. In vision, for instance, a molecule in an eye is struck by light, uncoils, and no longer fits its original slot. The light brings energy which loosens the atoms. The atoms ramble, and in the course of rambling their energy jostles away. The molecule remains frozen in its new and now incompatible shape. The ejection of the molecule from its slot lets another molecule change its shape, which triggers another reaction. That reaction triggers a pulse of current along the nerve to the brain itself. The nerve ramifies, the pulse is spread to a multitude of cells within the brain, and in each one its arrival results in a chemical modification. The cells' constitutions determine how they respond to future pulses, and whether they send new pulses down some channels or down others. And in due course, but perhaps not for a decade, the perception of an event influences a deed.

Every process in this chain is propagated from stage to stage through the motiveless agency of chaotic dispersal. The light loosens the molecule, which then uncoils by misadventure. The molecule is ejected because it has freedom to roam, and energy to lose. A reaction takes place when the ejection of the molecule allows the one remaining to explore new arrangements. The electric pulse is squeezed along the nerve by a sequence of reactions, each one triggered by its neighbour, and each trigger permitting molecules to wander into new arrangements. The chemical reactions in the ramification of nerve cells in the brain are similarly triggered, so is the deployment of the current as it rings for years through the brain. All the processes in the sequence, up to, including, and extending beyond the subsequent deed, are driven forward by the chaos they unleash. That subsequently we laugh or

We have to import material that undergoes reactions that release so much entropy that they can drive reactions in the body's cells back to their initial condition, ready to react forward again when next allowed to. As far as the nervous system is concerned, the principal agent is the *sodium pump*.[33] We have seen that the passage of a nervous impulse leaves the nerve with an exhausted distribution of sodium and potassium ion concentrations. The resuscitation is achieved with the sodium pump, which is a protein molecule embedded in the nerve wall. Each pump uses the energy available from an ATP molecule, a molecule of extreme importance in the reactions dealing with biological energy supply. When operating flat out, each pump can effect the exchange of around 200 sodium ions and 130 potassium ions per second. The array of pumps along the nerve restores the initial concentration difference, and sets it up ready to transmit the next pulse as soon as the cell body at its head has accumulated enough change of composition to induce it to fire again.

cry, or in due course love, argue, or despair, is determined by the long and complex history of events, all driven by dispersal.

I find it perplexing that to some, even now, it appears that the richness of the brain's properties, properties like perceiving, remembering, acting, deciding, and inventing, cannot have emerged by itself, or that such richness cannot be the outward display of inner motivelessness. It is so important to see through the illusion of complexity into the simplicity underneath. Of course, we might not be able to trace the simple steps that constitute a perception or an opinion or precede or bring about an action; but underneath there is no doubt that they are there. Yet I would not wish this view to be taken as an elimination of the wonder of life: it should, though, redirect the wonder. What wonder there is, should, in my view, not be at the benevolence and subtlety of external intervention, for that leads to the unnecessary intrusion of a spirit and the invention of a soul. It should instead be wonder at the realization that underlying simplicity can have such glorious manifestations when elaborately coordinated, and that such coordination can grow through the selection of evolution. The only immortal soul man has is the lasting impression he makes on other men's minds.

We do not see one thing and die. The body must be rewound to respond again, and nerves have to be prepared to propagate again. Every step in perception and action is reaction, and every reaction can be undone. A suitable reagent has to be supplied which has the energy to permit yet another haphazard exploration of arrangements, and to induce the molecules to re-form their earlier structures. In other words, we have to eat.

We have seen that perception and action are both powered by the tendency of energy towards chaos. Both degrade the quality of energy in the universe, and both ultimately involve energy's corruption. In eating we import more high-quality energy from our surroundings, and recharge our bodies by letting it disperse into our cells, where it is ready for the next step of degradation, such as an act of perception, of action, or of invention. Every action is corruption; and every restoration contributes to degradation.

At the deepest level, decisions are adjustments of the dispo-

The learning process can be investigated at a microscopic level in terms of the way that the synapses of the brain are modified by the brain's own activity. The emergence into the world of the underlying processes in terms of actions and emotions is described in a variety of places.[34, 35, 36] Detailed analysis and experiments on the small groups of neurons responsible for the activity of gill retraction (an action I refer to simply because it has been studied) are available[37]. The extraordinary feature of this kind of work (on a snail) is that the neuronal system studied is so simple that it can be taken apart like the components of a computer, and the role of individual components and their wiring diagram analysed just like any network of switches and wires.

We start (after the age of about 10 years) with about 10^{11} neurons. We do not renew them as they die (which is when for some reason their reactions cease). It is said that every tot of whisky eliminates 5000 neurons in addition to the 50 000 due to die that day.

sitions of atoms in the molecules inside large numbers of cells in the brain. The underlying reasons for those changes are the same as for any process. The atoms have no will to move, but given an opportunity they explore, with the risk of being trapped when the energy jostles into the world and dissipates. Every change in the complexion of the cells and their interconnections is at heart brought about by a natural disposition to chaos. That this motiveless, purposeless, mindless, activity emerges into the world as motive and purpose, and constitutes a mind, is wholly due to the complexity of its organization. As symphonies are ultimately coordinated motions of atoms, so consciousness emerges from chaos.

Decisions are taken on the basis of the predisposition of the brain. The manner in which chaos emerges into the world to take the name of action depends upon the state of preparation of its cells. The consequences of changing the state of a single cell depend upon the state already existing in the cells with which it is in contact. So the whole of our personal history, so long as our cells survive, channels the ramifications of chaos. That cells switch the activity of the brain to some cells and not to others, and in the process become modified so that subsequent pulses, perhaps returning from those recently stimulated, are channelled elsewhere, is the complexity of organization that feeding on chaos marshals it into coherence.

Inheritance, genetic information transmitted by reactions, lays down the structure of the brain and imposes a pattern of switches. Experience, the lifelong sequence of reactions responding to influence, then ceaselessly modifies and develops the network. Age is the death of cells and consequently the loss of subtlety. Senility is the decay of the sophistication of the organization of the circuitry, and the consequent failure of the brain to coordinate underlying chaos into brilliance.

So long as we can restore our cells by hunting for high quality, undispersed energy in the outside world, transferring some of it to our cells, then so long can our complexity be sustained. A bleak yet honest view is that living is therefore a struggle (a struggle ultimately driven not by purpose but by dispersal) to discard

37

low-quality energy into the surroundings and to absorb high-quality energy from them. In a sense, we corrupt the outside world in order to have an inner life. The chain of consumption, men eating cows, cows eating grass, grass eating mountains and living off the sun, is what has grown up through evolution as an interlocked mechanism of dispersal. There is no need to look for a purpose behind it all: energy has just gone on spreading, and the spreading has happened to generate elephants and enthralling opinions.

I should like to add a postscript. The singular property of the brain is that its response to circumstance is to a degree under its own control. It can take advantage of opportunities to select paths towards its own annihilation, as in despair or an inclination to martyrdom. Or it can take advantage of opportunities to select paths towards the fulfilment of its potential, as in acts of comprehension and creation. These inclinations are consequences of the pre-existing state of the brain, its chemical composition when the thought or inclination emerges and is rendered into action. Free will is merely the ability to decide, and the ability to decide is nothing other than the organized interplay of shifts of atoms responding to freedom as chance first endows them with energy to explore, and then traps them in new arrangements as their energy leaps naturally and randomly away. Even free will is ultimately corruption.

Second Orientation

We have seen that the motivation of all change is the natural dispersal of energy, its spontaneous collapse into chaos. The richness of the world, the emergence into it of art and artifacts, of opinions and theories, can be traced down to the level where it can be seen to be merely the gearing together of steps to dispersal. Acts of creation are temporary and local abatements of chaos, but every abatement is driven by a greater storm of chaos elsewhere. All change, change in all its forms, is the outcome of a complex web of interconnections which happen to channel the unwinding of the universe into discernible events. Underneath opinions and deeds there is nothing to explain beyond the unravelling of the network of connections that transforms the natural and the understandable into the unexpected and the imaginative. Ultimately there is only chaos, not purpose.

We now seek the nature of the ultimate jostling, and investigate what governs the places where atoms go and energy leaps in the universal interconnected unwinding. I shall try to identify what determines the motion of individual elementary things, and account for the rules that seem to govern their behaviour. I shall argue that the natures of things determine their destinies. This exposes more of the infinite laziness of the lazy creator, because it shows how allowing total freedom leads to the emergence of constraints. I shall attempt to show that doing nothing, except allowing absolute freedom, leads to apparently regulated behaviour.

We shall see how noticing the obvious lets us discern the underlying nature of light and matter. We shall come close to discovering the nature of space and time and seeing their amalgamation into spacetime. We shall see that the fundamental natures of energy, force, and particles emerge as properties of space and time. In this way the nature and singular properties of time begin to become comprehensible, and the world a little clearer.

Three

HOW THINGS CHANGE

To understand the terms 'jostle' and 'packet of energy', we have to examine the behaviour of individual entities, such as atoms and molecules. We have to see what determines the direction of flight of a molecule in a gas, and what determines how the direction changes when there is a collision. These are problems of mechanics. There are two great systems of mechanics: *classical mechanics* and *quantum mechanics*. The former was a great achievement of the intellect (Newton's principally) and essentially initiated theoretical physics. Yet it is only an approximation, and quantum mechanics has displaced it as the most fundamental description of the properties of matter currently available. We shall see their interplay.

THE POINT I NOW WANT TO MAKE IS THAT THE BEHAVIOUR OF things is determined by their nature. I want to show that the essential quality of very simple things, even mindless things, is sufficient to determine their behaviour without it being necessary to impose any rules. Things as small as atoms and electrons cannot take decisions, but have to behave in accord with their innate character: what things are, even on an atomic scale, determines how things are. An infinitely lazy creator would avoid the specification of rules if an entity's nature alone could govern its behaviour. We shall suppose that it does and he did, and shall show how it is possible to discard the laws that the world seems subject to. The only assumption we shall make is that things happen unless they are expressly forbidden; and nothing will be forbidden.

We have seen that the flow of events in the universe is the interlocking of steps towards the dispersal of energy. Individual jostlings are shifts of atoms, accelerations, collisions, and so on. In this chapter we look more closely at these steps, and examine how a species responds to and affects its neighbours. In Chapter 2 we moved towards simplicity by discarding purpose; in this we move further towards it by discarding the need to impose rules. Things behave, rules are our commentary on their behaviour.

One way of expressing our examination of the creation is that we are seeking the absolute nature of things. The attitude that intrinsic character governs observed behaviour is more useful in such a search if it is turned round. We shall take the view that we can discern the essential qualities of things by noting their behaviour. If we can identify a characteristic that unavoidably entails the behaviour observed, then we shall have identified something crucial, some essential aspect of the thing.

We shall therefore inspect the universe around us, select a few obvious things, and identify the simplest rules that capture their behaviour. The next step is then to look for the characteristic of the things which, once discerned, enables us to explain and in effect

45

Here, as so often in science, differences of kind give way to interpretation in terms of differences of degree. While light appears to propagate in a way quite different from sound, they are intrinsically the same, but show their similarities in a deeper way than might be suspected. This is a point to be developed in what follows.

The straight line propagation of light is an aspect of *Fermat's principle of least time*.[22, 38] We shall later encounter other examples of *extremum principles*, and see that this is a special case.

discard the rules. At that stage we shall have eliminated unnecessary complication.

The easiest place to start is with the observation that underlies our major mode of contact with the world and the universe beyond: the observation that we can see.

Light, we all know, travels in straight lines. If it could bend round corners, the world would be harder to discern. It would be like listening to it instead of seeing it. We would be immersed in a symphony of colour from objects that could be vaguely located but only hazily scrutinized. There would be no night; the symphony would be endless.

But saying that light travels in straight lines is not quite right. It conflicts with observation. Light bends at the junction of different media. Your leg in your bath looks broken even if it isn't. A lens bends light, and is shaped to focus the image on a film or on an eye. We therefore have to find a rule that captures both the straightness of the path when the medium is uniform and its bending when it passes from one medium to another.

The rule that captures both turns out to be elegantly simple (like all acceptable rules prior to their elimination): light travels by the path that takes the least time.

This succinct rule obviously accounts for the motion of light through air or any other uniform medium, because a straight line is then also the briefest path for anything travelling with a uniform speed. The rule also accounts for light's bending at the junction of media. Light travels at different speeds in different substances; the briefest path is then no longer the straightest, as can be understood by thinking about drowning.

Suppose the victim is out to sea, and you are on the shore. What path brings you to him in the shortest time, bearing in mind that you can run faster than you can swim? One possibility is for you to select a geometrically straight path from your deckchair to where he is sinking: that involves a certain amount of running and swimming. Alternatively you could run to a point on the water's edge directly opposite him and swim out straight from there. That is greater in distance but it may be briefer in duration if you can run very much faster than you can swim. By trial and error, or

This discussion is essentially Fermat's construction of Snell's law of refraction, that the ratio of the sine of the angle of incidence to the sine of the angle of refraction is equal to the ratio of velocities in the two media (and inversely proportional to the ratio of refractive indices). An engaging exercise is to calculate your actual trajectory from a specified position of your deckchair on the beach to a specified position of the victim. The complexity of this calculation (or at least, the time it takes to come to a numerical conclusion) underlines the fact that light must have if not a natural instinct then at least an intrinsic characteristic that guides its propagation.

A light wave is an electromagnetic wave, a series of peaks and troughs of electric and magnetic field.[39, 40] The wavelength of the wave determines the perceived colour. Visible light lies in the range of wavelength 4×10^{-7} m (violet) to 7×10^{-7} m (red). These wavelengths can almost be imagined: 10×10^{-7} m is one-thousandth of a millimetre, almost within the capture of the mind's eye.

Underlying these paragraphs is the concept of *interference*. Interference between waves may be either *constructive* or *destructive*. In constructive interference the displacements of various waves superimpose in such a way that they lead to an augmentation of the total displacement: the superimposition is of large amplitude. In contrast, in destructive interference the displacement interfere in a way that cancels, leading to a small total amplitude. This effect can be seen on the surface of a pond after two stones have been dropped in fairly close to each other: the concentric circles of displacement of water augment or cancel each other where they spread into a common region.

trigonometry, you would find that the path involving the least time is one where you run at some angle across the beach, then change direction and swim at another angle in a straight line towards your target (if it is not too late by now). This is exactly the behaviour of light passing into a denser medium.

But how does light know, apparently in advance, which is the briefest path? And, anyway, why should it care? The only way of discovering the briefest path appears to be to try them all, and then to eliminate all traces of having done so. There must be something about the nature of light which entails that it naturally tries all paths, and then eliminates all but the briefest.

The essential property is that light travels as a wave. Once that is realized, its other properties fall into place: light cannot help travelling by the briefest path.

A wave is an undulation, a series of peaks and troughs. Two or more waves of disturbance may spread into the same region. If it happens that the peaks of one coincide with the troughs of the other, then they tend to cancel, and an observer sees less disturbance, and perhaps no disturbance at all if they happen to cancel completely. That is basically all the information we need in order to see how light's character determines its destiny.

We are taking the view that things happen if they are not expressly forbidden, and that an infinitely lazy creator does not trouble to forbid. Think, then, of a light ray that happens to travel from A to B along some meandering path. *We* know that light doesn't travel like that, but light doesn't. If that path is permissible, then so too, as far as the light is concerned, is one that lies very close to it. So the light also travels by that path. Whereas the light that snaked by the first path may have reached B with a peak, the light that snakes by the second might reach B with a trough, or something in between. There are very many paths lying close to the first, and an observer at B sees the total disturbance arising from the waves that explore them all: many are troughs at B, many are peaks, and many are all the possibilities in between. The total disturbance at B is consequently zero, because there is always a neighbour to wash its neighbour out. In other words, by letting light travel by any path, it appears to be unable to travel at all. But light does travel.

When the wavelength is short, even small changes of path can cause dramatic changes in the interference. In other words, short wavelengths act as a kind of magnifying glass and magnify small displacements of path into large changes in the extent of interference, as a result, only paths very close to a straight line escape destructive interference when the wavelength is short. Conversely, when the wavelength is long, the interference depends only feebly on the path.

The wavelength of middle C is 1.3 m, comparable to the size of typical obstacles in a room. Sound is a pressure wave, not an electromagnetic wave; but the considerations applied to the propagation of light apply (except in matters of detail) to the propagation of sound.

The slowing of light when it passes into a denser medium is a result of the interaction between the electromagnetic field of the light ray and the electrons of the molecules forming the medium. In a sense we have to build this into the argument as well as the wave nature of the light; but it takes care of itself when the universe is looked at as a whole. A lens can be regarded as a region of dense medium especially shaped so as to induce the correct interference between rays of light that pass through, and so to bend the rays to an appropriate focus.

One step was too hasty. Think of a ray that happens to go straight from A to B. Now think of a neighbouring path and the ray that takes it. If that path lies close to the first, it will have a trough at B if the first had a trough, and a peak if the first had a peak. There are very many almost-straight lines from A to B, and they all give disturbances at B differing only slightly from the disturbance due to the straight path. These paths therefore do not wash each other out, and an observer at B sees the light. He observes that the light travelled to him by lines that are straight, or very nearly straight.

The extent to which the nearly but not quite straight rays contribute to the overall disturbance at B depends on their wavelength (the separation of successive peaks). If the wavelength is short, then only rays correspondingly close to the straight line survive, all the others having sufficiently destructive neighbours. As the wavelength increases the waves get out of step less quickly, and the eliminating power of neighbours declines. Then even quite bent paths survive and can deliver their disturbance. That is the reason why radio transmissions (which use long-wavelength waves) can circumvent houses, and why we cannot see round corners. We can hear round corners: sound waves' wavelengths are long.

The wave nature of light accounts for the inevitability of its selection of straight lines. That, though, is true of uniform media, as in air. When light passes from one medium into something denser it travels more slowly. As a result, the positions of its peaks and troughs are modified. Still it explores all possible paths, but no longer is the geometrically straight line the one without neighbours that annihilate. Now the surviving path, because of the shift of peaks and troughs, is the one that bends at the junction. The surviving path also happens to be the briefest path. That rule therefore turns out to be merely a distant commentary on deeper purposelessness. Light automatically discovers briefest paths by trying all paths, and automatically eradicates all traces of its explorations; this presents itself to us as a behaviour, which we summarize as a rule.

We see in this example how perfect freedom generates its own constraint. As well as accounting for observed behaviour,

The principle of least time is an example of an 'extremum principle'. The classical mechanics of particles can also be expressed in terms of an extremum principle, the *principle of least action*. Maupertuis, the originator of the principle (in a somewhat confused form and in 1744), put it forward in an attempt to provide a theological foundation for mechanics, the argument (as so many that have been spawned by minimum principles) being that the perfection of the Supreme Being would be incompatible with anything other than the minimum expenditure of action.[22, 38] I find it engaging to see how the tail of this argument has now wagged the dog out of existence.

everything we have said accords with the common-sense view that inanimate things are innately simple. That is one more step along the path to the view that animate things, being innately inanimate, are innately simple too.

The next step in the development involves noticing a similar observation about another thing. Since the behaviour is similar, we can suspect that the explanation is similar too. I should like you to notice that particles of matter also travel in straight lines unless subjected to a force. Why?

According to the view we are taking, they do so because it is their intrinsic nature. But what can be this intrinsic nature that determines such behaviour? It must be that particles are distributed as waves.

In a single leap, impelled by common sense, we have gone from the old-fashioned original physics of Newton to the modern theory of matter, quantum theory, which regards the qualities of 'particle' and 'wave' as inseparable. Many feel at home with classical physics and regard quantum theory, being less familiar, as contrary to common sense. In my view, though, common sense drives us to accept quantum theory in place of classical physics as more consistent with common sense. I hold that the mind-shutting familiarities of classical physics actually conceal its incomprehensibility, except as a commentary and a mode of calculation. When they are inspected, the explanations of classical physics fall apart, and are seen to be mere superficial delusions, like film-sets.

There is much more to quantum theory than the assertion that particles are intrinsically wavelike, but that remark is at its core, and is what we develop here, first by seeking the actual rule that appears to govern the classical mechanics of particles, and then by looking for an explanation.

The rule that appears to govern the propagation of particles is remarkably, and therefore suspiciously, like the rule that appears to govern the propagation of light: particles follow trajectories between A and B that involve least action. Never mind the technical meaning of action; it is good enough, and truthful enough, to think of action as having its everyday meaning. In particular, if the particle is not subject to a force then the path that involves least action—no meandering and no acceleration—is uniform and straight.

The switch from classical mechanics to quantum mechanics in terms of the analogies between the principle of least time for optics and the principle of least action for mechanics is established in detail by R. P. Feynman.[41, 42]

The straight lines of warped spacetime are its *geodesics*. Now we are moving into the domain of general relativity.[43–37]

. Particles propagate along geodesics in spacetime for the same reason as they propagate along straight lines in ordinary (but non-existent!) Euclidean space. Note that a timelike geodesic in spacetime is locally the *longest*, not the shortest distance; nevertheless the same arguments apply.

Now, how does a particle know, before it tries, which of the infinity of possible paths from A to B corresponds to the one of least action? And why should it care?

As soon as we take the view that particles are distributed like waves, both questions are eliminated by the same reasoning that eliminates them in the case of light. The intrinsic nature of particles, their wavelike character, ensures that they travel in straight lines of least action, because all other paths, which they are perfectly free to explore, are eliminated automatically. The reason why particles like pigs and people do not normally seem to be waves is simply that their wavelengths are normally so short as to be undetectable. Nevertheless, distributed as waves they are, and that attribute provides explanations which are totally beyond the reach of classical physics.

This picture accounts for motion in straight lines, because in the absence of forces such paths are survivors; the wave nature lets them survive. On the other hand we know that curved paths, and accelerated motion, arise from the action of forces. But what are forces? As a first step towards finding an answer we look at the nature of gravity, for gravitation builds the stage on which other forces play out their roles.

Gravity, in brief, is warped spacetime. With that understood (and in a moment we shall set about its understanding) gravitation as a separate concept is effectively eliminated, except from conversations and calculations. Elimination is the best explanation, for it implies less creation.

For the moment forget the time in spacetime and think about the space. Then warped space means that what we perceive as straight lines are different from what are truly straight lines. In a sense, it is our perception that is bent, and familiarity is responsible for yet another illusion. Therefore instead of explaining why particles turn from straight lines under gravity, we shall adopt an elegant alternative. We shall take the view that particles invariably follow straight lines, but those paths look bent to us.

This change of viewpoint, although it appears to be a philosopher's delight of a play on words, is in fact a great simplification. There is nothing slippery in a change of viewpoint: science is the

The gravitational field of the sun takes exactly one year to make a straight line appear to be a circle the diameter of the Earth's orbit.

The nature of time has, of course, been subject to much discussion, which has variously been theological, philosophical, and fruitful. The most interesting recent account seems to me to be the book by G. J. Whitrow[48] which deals with various levels of the aspects of time, among them human time, biological time, mathematical time, and cosmic time. This fascinating book ranges onwards from jet-lag in bees. Other interesting compilations include the books by Denbigh[49], Coveney and Highfield[50] and the collection of lectures edited by Flood and Lockwood.[51]

search for the simplest view—the view that eliminates elaboration, the view that does away with further questions. Now we have one less thing to explain, because we know already that particles have an intrinsic nature that leads them to pursue straight lines. Things—particles, light, people, planets, and stars—*invariably* follow straight lines. That is their nature. We, the observer and commentator, however, have to modify our man-made conception of straightness.

Nevertheless, warping space does not seem quite to do the trick. For instance, we know that planets travel in more or less circular orbits around the sun, and it doesn't seem possible to imagine warping space to the point of twisting straight lines so that they look like circles. Moreover, we also know that under the influence of gravity things accelerate. In order to think about speed, we have to think about time—position can exist without time, but speed needs time. Once time is brought into the discussion, and space and time considered jointly as coalesced into spacetime, it not only brings in speed but provides enough flexibility to distort straight lines into apparently spatially closed orbits. In other words, observing that planets and satellites travel in approximate circles compels us to think of space and time together. Noticing the obvious has led to another synthesis.

But what do we mean by the coalescence of space and time? The amalgamation must be a procedure of some delicacy for they seem such disparate entities. We can see spatial extent, walk round it, turn it over, and then look at it again; but we seem to be merely conscious of time, and are incapable of arresting it. Whereas we can grip objects in space, it is time that grips us. Time seems internal, space external.

Time is distinct from space even in the absence of consciousness. While people refer to it as a fourth dimension, it is not simply a fourth dimension of space. While we might think of time as being a single extra dimension, spacetime is not merely one more dimension tacked on to space. The essential distinction between space and time lies in the nature of their amalgamation. It lies, in a word, in their geometry.

The richness of existence boils down to a geometrically peculiar-

The idea of measuring distances N–S and E–W in different units as an analogy for the incorporation of time is a development of an idea that has been taken further.[52] The deeper advantage of the conversion is that it leads to the concept and definition of 'distance'.

The precise value of c is $2.997\,925 \times 10^8$ m s^{-1} ($186\,282.4$ miles s^{-1}). The reason why I say it is misleading to refer to it as the speed of light is that it is the limiting speed for the propagation of any kind of signal (or object); it merely happened to enter physics *via* electromagnetic theory. Thus any signal conveyed by massless particles (such as photons for electromagnetism, gravitons for gravitation, and neutrinos) is propagated with the same speed c.

A common measure of distance is the light-year. One view of its significance is that it is the distance travelled by light in one year (and corresponds to 9.45×10^{15} m, or 5.87×10^{12} miles, 5.87 million million miles). The sun is 8.3 light-minutes away, the

ity. Its nature can be discerned by means of the following analogy. Think first of space alone. Suppose that by an accident of history we were in the habit of measuring East–West distances in miles and North–South distances in kilometres. Think how cumbersome it would be to express the distance (and to express the time it would take to drive) from here to a point that lay to the North-East! There would also be great perplexity when people found that the radius of a circle varied from 1 mile to 1.6 kilometres. Surely, they would have thought, there must be a clue to the nature of the universe in that observation.

The complexity would have vanished when someone suggested converting miles to kilometres. As soon as all East–West distances are converted to kilometres, the radius of a circle is found to be independent of direction, North–East distances and all others take on a very simple form, and Pythagoras could express his theorem, that $d^2 = x^2 + y^2$, with great simplicity. The complexity of the geometrical description of the world drops away when all separations, in whatever direction, are expressed in the same units.

You can imagine that there would have remained scientists, and not a few philosophers, who would go on trying to explain the source of the magical conversion factor from miles to kilometres, believing that its value, 1.609 344 km per mile, held a central place among the attributes of the universe in need of explanation. We should not get side-tracked into seeking insight from the man-imposed but seek it instead from their elimination.

Exactly the same kind of simplification occurs when the units of time are brought into line with the units of space. All we need is a factor that converts from seconds to kilometres, a factor of so many kilometres per second; that is, a speed. The factor that works, in the sense of agreeing with observation, has the value of about 300 000 kilometres per second, a quantity denoted c and previously, and misleadingly still, called the speed of light.

When time is expressed in kilometres there are many everyday inconveniences, such as the hugeness of the numbers on the faces of clocks, and unfamiliar infelicities such as 'You are 56 million kilometres late', but the rich reward lies in the elegant simplicity of the measurement of separations and the description of curves.

nearest star, *Proxima Centauri*, about 4.27 light-years, and the nearest galaxy, the Andromeda Nebula, about 2.25 million light-years. Another view of the light-year is that it is the outcome of expressing distances in units common with those of time (i.e., doing the opposite of what we have been mentioning so far: it is just as legal and just as sensible to adjust man-made units in one direction as in the other).

The observation that the speed of light is independent of the state of motion of the observer is the celebrated Michelson–Morley experiment, which gave one of the great negative results of science.[48, 53] The other great non-event was the failure to observe differences between inertial mass and gravitational mass, as in the experiments of Eötvos and Roll, Krotkov, and Dicke. This failure can be regarded as the spring of general relativity theory.[11, 44]

We shall refer to the signs $(+,+,+,-)$ in $(+)x^2 + y^2 + z^2 - (ct)^2$ as the *metric signature* of the space. The signature of Euclidean geometry in four space dimensions would be $(+,+,+,+)$.

The *distance*, the *d* in $d^2 = x^2 + y^2$, between two points was introduced as a quantity characterizing space. The *interval* between events is the quantity characterizing spacetime. It is found that the form $x^2 + y^2 + z^2 - (ct)^2$ does not depend on how fast the observer is travelling.[52]

With c merely a conversion factor for man-imposed measurements, the puzzle is really why light happens to travel with a speed that has the same numerical value. The answer will emerge as we pursue another feature.

As c is a factor that converts one unit of measurement into another, it is plausible that every observer will ascribe to it the same value. It shouldn't matter where you are or what you are doing, the conversion factor ought to be the same. Every observer, irrespective of his state of motion, should measure the same value of c.

Incorporating the constancy of c into physics led to the second of the twentieth-century's scientific revolutions: relativity. That revolution emerged from a conflict. On the one hand c is identified as a relative speed (the speed of light relative to an observer, who may himself also be a traveller and therefore reasonably expect to measure different speeds of light depending on whether he was approaching the source or receding from it). On the other hand c is identified as an absolute constant, fixed for every observer whatever he is doing and, in particular, how fast he is travelling.

There is only one way to resolve the conflict. Time has to be attached to space in such a way that it distorts our perception of relative speeds.

Pythagoras' theorem can be summarized by the expression $d^2 = x^2 + y^2$. Its extension to separations in three dimensions is $d^2 = x^2 + y^2 + z^2$. If time were merely a fourth dimension of space, a modern Pythagoras would write $d^2 = x^2 + y^2 + z^2 + (ct)^2$. That extension, though, simply doesn't work. An observer measuring the separation of events according to this expression would find himself in a hopeless muddle, with the separation depending in a complicated way on his speed, and with his measurement of the speed of light depending on his state of motion.

One small change eliminates this confusion and results in all observers observing that light travels at the same speed, whatever their own states of motion. If separations are given by $d^2 = x^2 + y^2 + z^2 - (ct)^2$, then they become independent of the observer's activity and orientation. Every observer then records the same separation. Every observer reports the same speed of light.

The restriction on the direction of signals arising from the metric signature $(+,+,+,-)$ will be found discussed in books on relativity.[52, 54] See also the reflections on time by Prigogine.[55]

As an illustration of what is and what is not a straight line in spacetime, consider your own trajectory as you sit reading this book. The gravitational influence of the Earth distorts spacetime in your vicinity in such a way that your natural direction of motion is towards the centre of Earth. While tending to pursue that direction you encounter an obstruction, the Earth itself (which you cannot penetrate for quantum mechanical reasons, unless you shovel it aside). This exerts a force, which you can experience through the seat of the chair. That force is deflecting you from your geodesic.

We have come upon the crucial simplification of the description of the properties of space and time. It is on account of that minus sign in the expression for the separation that time is not merely a fourth dimension of space even if it is expressed as a distance. It is also on account of that minus sign that time is quite distinct from space. We shall see that that apparently minor change of sign is the difference between existence and non-existence, as well as at the root of our different perceptions of location and duration. That minus sign is what we mean by a geometrical peculiarity. It underlies both the existence and the evolution of the universe.

On account of the minus sign no signal can penetrate backwards through spacetime any more than a common circle in familiar geometry can have a negative radius. The minus sign insulates the past from the present and ensures that nothing in the future can modify something now or in the past. It ensures, therefore, that our destinies lie in the future and not in the past.

The unrepeatable, irreversible sequence of events that constitutes our consciousness is the 'time' we perceive as flowing forwards. Those events necessarily pace forwards along the dimension we refer to as time. As individual events project only forwards in time, and as change is trapped into irreversibility by dispersal, and as perception is the accumulation of experience, so our consciousness is carried into the future.

We have to return to the illusion of straightness and the natural trajectories of particles. We saw what happened when space alone is warped; now think of spacetime as warped. Spacetime is twisted by the presence of matter so that what is apparently a straight line is not in fact intrinsically straight. Now, particles pursue straight lines through spacetime because that is their nature; but a straight line is no longer perceived as straight by an onlooker. Because time is now also folded into space, and the overall structure is twisted, uniform motion is no longer perceived as uniform any more than spatially straight is perceived as straight. Instead, an onlooker perceives accelerations and decelerations of the particle. The motion of the particle actually remains uniform, but the intrinsic properties of spacetime deceive the onlooker into perceiving modifications of motion.

Probably the most readable introductory account of the interpretation of forces in terms of the exchange of particles will be found in the book by Nigel Calder[10]. The comment about boomerangs was picked up from a 1980 Wolfson College lecture by Sir Denys Wilkinson. The analogy is deep as well as graphic, for whether or not forces act attractively or repulsively depends on the spin of the particles being exchanged: even-integral spin particles are attractive, odd-integral spin particles are repulsive between like particles. For a survey of the modern description of forces see Davies[56, 57] The unification of the forces and the concept of supergravity has also been described quite simply.[58]

The electromagnetic force between electrically charged particles is due to the interchange of *photons* (light quanta). The strong force is mediated by *gluons* acting between quarks. The weak force is mediated by the unhappily but accurately named *intermediate vector bosons*. The gravitational force arises from the exchange of *gravitons*.

The path of a planet around the sun is a line that is perfectly straight and traversed uniformly, but we perceive it as a closed and varying orbit. The rise and fall of a ball is actually a straight and uniform path, but the twist induced in spacetime by the neighbouring Earth distorts our perception like an inferior lens, and we infer that it is influenced by a force that retards the ball and then brings it back. The force is not actually there: the trajectory is an illusion.

Gravitation is the word we use to signify this distortion. Motion is intrinsically extremely simple: motion is uniform and straight, it merely seems distorted by the presence of matter. There is at root no such thing as a gravitational force.

The obvious question now is why matter induces distortion: I prefer to set this aside for now and return to it later. There remain questions to do with the main themes. There are forces other than gravitation: there are electrical forces, and still others. What is their nature?

An analogy will help us first to identify and then to eliminate their existence. Think of a pair of ice-skaters skating parallel to each other. When they start to throw balls to each other, they drift apart as they throw their own and catch the other's. A distant observer, not seeing the balls, will easily be deceived into believing that one skater is repelling the other. He will infer that there is a force between them, a force that drives them from their straight paths. We know better. We know that they are exchanging things. If the skaters throw boomerangs, a distant observer sees them drift together. Since he does not see the boomerangs, he infers that there is a force of attraction between them. We know better. We know that there is no such thing as force, merely an exchange of things.

All the forces that bind atoms, nuclei, and the deepest components of particles, can be regarded as arising from the interchange of particles. Force is only the codeword for this behaviour being played out on the arena of spacetime. Spacetime, with its curvature forms the stage; particles distributed as waves pursue straight lines; but particles detach from particles and travel (straight) to others and impress on them their motion. Force is the name of this interchange of particles. There is nothing else to force.

Third Orientation

We have seen how nature governs destiny, how behaviour reveals nature, and how perfect freedom generates its own constraints. We have seen that light and particles are both distributed as waves. These waves explore with total freedom, and in eliminating their traces delude us into regarding things as obeying rules as well as having natures. The peculiar role of gravity switched our attention to the nature of space and time, and we saw that perfect freedom still accounts for observed behaviour as soon as space is amalgamated with time, and spacetime is distorted so that straight lines are no longer perceived as straight. Even apparently complex motion, such as occurs under the influence of other forces, emerges from perfect freedom when the nature of forces is understood as the interchange of other particles. We also saw that the amalgamation of time and space is done in a way that happens to ensure the insulation of the past from the present, but not the present from the past. Our progress through eternity, or what there is of it, is an aspect of geometry.

Since spacetime is so important we need to examine it more closely. In particular we should examine one of its most obvious characteristics, its dimensionality. In the chapter that follows we turn to an inspection of the connection between the existence on the one hand of three dimensions of space and one dimension of time and, on the other, of consciousness able to discern and respond to them. We also begin to see why, when a universe comes into existence, it adopts our familiar dimensionality. So we are brought to the heart of the nature of matter and energy, and we come to the edge of seeing what the infinitely lazy creator must bring into being (or, being uninvolved, not stop from being).

Four

WHERE THINGS CHANGE

We are in the realm of self-reference, a labyrinthine and self-absorbing subject. By far the most extraordinary portrayal of self-reference in literature, music, art, and mathematics is D. R. Hofstadter's delightful book.[59] It shows how witty spacetime can be when it is looked at from the point of view of self-reference.

The difficulty of imagining spaces of unfamiliar dimensionalities should not stop us contemplating them: mathematics lets us discuss them (which is either an extraordinary aspect of the power of mathematics, or an aspect of its powers of self-reference). Four-dimensional objects can be portrayed in three dimensions (just as three-dimensional objects are portrayed in two-dimensional pictures). Hofstadter, in the book mentioned above, gives a reference to the construction of an optical illusion for four-dimensional people.[60]

WE LIVE WHERE THERE IS UP AND DOWN, SIDE TO SIDE, AND backwards and forwards. The obvious thing about the universe is that it has three dimensions. Why is that so? Why doesn't it have only two dimensions, or four, or even more?

Through questions as fundamental as these we can begin to understand not only the phenomena of nature but also the processes of creation. The space we inhabit is more than a stage for events, because matter itself is space (as we shall see more clearly later). Therefore, the creation can be regarded as merely the making of space.

Space, you might think, is a simple thing to create; but it has to be something other than mere vacancy if it is to have properties as rich as those that constitute the perceived world. I shall attempt to justify the view that, in a sense I wish to be taken as being wholly devoid of overtones of mysticism, space itself is self-conscious.

Have you also ever wondered why there is only one date on your daily newspaper? Why, in other words, is there only one dimension of time. Why isn't time two-dimensional, or even more? Would our perception of time be intrinsically different if it were two-dimensional, so that we could move sideways in time (to a different time) as well as forwards?

The geometry of space lets elsewhere be accessible from anywhere; but the geometry of spacetime insulates the past from the present. Does the past's inaccessibility depend on time's dimensionality? If so, then perception is a feature of dimension.

We shall explore the nature and consequences of the dimensionalities of space and time. We shall begin by examining what kinds of universes might have emerged if the creation had generated spaces with different numbers of dimensions. Later we shall see that it is quite probable that the coming about of creation could only result in the formation of our familiar universe with its three dimensions of space and one (in a sense only half) a dimension of time. Nevertheless, there are advantages in considering other conceivable universes because they highlight the unique and possibly

The brain is a nonlinear transmission line. Information enters it from one side (the senses) and it transmits it to some kind of generalized receiver (an action or a statement). For an input to give a different and more complex signal at the receiver one needs a highly structured, nonlinear line—contrast the brain, for instance, with a copper wire, which is also a transmission line but is too simple to develop an input into an opinion of its own.

unsuspected characteristics of the familiar. The exploration will also uncover entertaining consequences of dimensionality, such as its implications for the nature of evolution and the emergence of consciousness.

In a nutshell, I shall argue that not only is a universe with three dimensions of space and one of time the only kind that can survive its own creation, but also that such a universe is the only one that has the capability of becoming self-aware.

I begin with self-awareness. While consciousness is plainly an artificial criterion for the viability of a universe, it is undeniably a property of this one. We might take the view that we are merely aware of *this* creation, and that there are other profoundly different universes that have stemmed from other creations in other places. If so, we ought to be aware of what these profound differences would entail. While pointing to them I shall argue that ours is the simplest and possibly the only type of universe (in the sense of having the right dimensionality) that can be aware of these other possibilities (as well as being the only type of universe that can be aware of itself). Later even the possibility of these alternatives is dismissed. At that stage we shall see that the pre-eminence arrogated by man turns out to be the pre-eminence of his underlying dimensionality.

First consider space. We shall examine the nature of universes with different spatial dimensionalities in terms of the consciousnesses they might support. If a component of a universe is to perceive, assimilate, learn, and communicate, it has to be complex. Consciousness is simply complexity. Being a being means organizing responses, and marshalling coherence out of the potency of chaos. We shall develop the argument on the basis that complexity of construction is the soul of being conscious.

A universe with no dimensions is a featureless point. As such it has no properties, no complexity, and certainly no self-awareness. Lacking dimensions implies lacking existence.

A one-dimensional universe is a line without width. The structures inhabiting it would be infinitely thin needles lying along the line. There is no up and down, and no side to side, only backwards and forwards. The needles can never pass, except by losing

A one-dimensional being might show intelligence if it developed a system of interacting nonlinear waves—the nonlinearity might introduce the complexity necessary for intelligence, but assembling the nonlinearity might be beyond such a universe.

There are a number of accounts of life in different dimensions. The immeasurably greater richness of the two-dimensional world is well exemplified by an article by Martin Gardner[61] based on what he terms 'a 97-page tour de force' by A. K. Dewdnay[62, 63] on two-dimensional science and technology'. The article and the booklet survey the science, the technology, the chemistry, the weather and the art, as well as the art of living, in a planar universe. A very helpful survey of the physics of an N-dimensional universe, and many further references, is given by Barrow and Tipler.[22]

The point about the gut is a development of an idea described by Whitrow.[64] Dewdnay's article on the planiverse[62, 63] discussed similar points of biological infelicity (or felicity in some cases; such as the difficulty of falling over and the ease of mowing lawns).

their identities and merging with their neighbours. Evolution depends on the incorporation of whatever happens to be found at either of their ends. Because a creation generates only simplicities, the things the needles encounter are always inanimate and no complex organic structure could ever evolve. Communication could develop, because, for instance, a needle could nudge its neighbour and that neighbour could nudge its. In time these nudges might coordinate into complex, waves of pressure rippling backwards and forwards along the universe. The waves, though, would no more be consciousness than atomic vibrations are bright ideas. On account of the poverty of the interconnection of the components, there is no opportunity for the development of the ability to learn, and the needles would have only the most rudimentary perception of the world.

In contrast, a two-dimensional world would be immeasurably richer. For instance, you could avoid your neighbours. You could also acquire new neighbours to enliven reproducing, and you could go hunting to enliven digesting. You would go round obstacles instead of forever being confined by the first you met. You could sustain your own complexity (at the expense of your surroundings) by acquiring and discarding food, and you could develop it (at the expense of your competitors) by encountering partners to collaborate with in evolution.

Nevertheless, there would remain infelicities. An amusing one relates to eating. If the efficient way of incorporating pieces of the outside world is through a gut, using one end to devour, the other to discard, and the rest to digest, then you would encounter severe structural problems in two dimensions. A two-dimensional you with a gut is two yous; you would have either an identity or a digestion. One way round this dilemma would be to dine in collaboration, each partner providing one side of the gut. Alternatively the gut could be a cul-de-sac, with the discarding done at the same end as the devouring. Such points, however merry, are more questions of social conventions at dinner parties than deep problems of the nature of the world; but they do indicate that dimension has an impact on etiquette.

Much more pertinent is the observation that subtle conscious-

The complication of a two-dimension brain[65] has been disputed[61] on the grounds that nerves may be able to develop a way of firing through an intersection; the brains would merely operate more slowly. The proponent's, Whitrow's, argument carries more force in practice even though the opponent, Dewdnay, is right in principle. Note that whereas the human brain has around 10^{11} neurons the number of synapses is much greater, perhaps around 10^{14}. That is a measure of its complexity.

The minimum size of a human planar brain would appear to be about 36 m²: this figure assumes that there are 10^{11} neurons in a sphere of radius 5 cm in the human head, and then all neurons are laid side by side in a densely covered surface. This, however, is simply a flattening of the head, as might be achieved with a hammer; its reconstruction with avoided crossings in two dimensions would lead to a vast structure, and dog-town seems quite reasonable as an indication of size.

Recall that a neuron fires when it has accumulated a net firing signal, this being composed of numerous inhibiting and stimulating signals arriving at the synapses.[33] The brain is a temporal as well as a spatial array.

ness could not emerge. For one thing, there would not be time for it to evolve, for another it would be stupid if it did, and for a third it would perceive its surroundings only dimly, and be confined to introspection by confused communications with others.

Consciousness, we presume, depends on complexity, the existence of a network of interconnections between cells elaborate enough to constitute a brain. A problem immediately arises in two dimensions. Whereas it is true that connections can lie side by side and circumvent obstacles, because there is neither up nor down they cannot cross. As a consequence, a single fibre may have to thread round obstacles for great distances before it can reach its destination. Some destinations may already be surrounded by cells and their appendages, making them wholly inaccessible and so diminishing the brain's opportunities for complexity.

There are many dismal consequences. In the first place, two-dimensional brains need to be immense. Instead of cells being connected by fibres a few centimetres long, as inside our heads, the connections may have to wind through several kilometres in order to achieve a network of sufficient complexity. The brain of a dog might be the size of a town.

Extent brings its own complication. The performance of brains depends not only upon the spatial arrangement of connections but also on the timing of signals. Although spatially complex networks might just evolve, they would fail to be viable unless even more complexity, in the form of holding areas and stacks, were introduced in order to regulate the arrival times of pulses.

Brains must be both drained and sustained. Somehow every cell in a flat brain has to be supplied with fuel, and the waste removed after the quality of energy has been transferred. Not only do the nerve connections themselves have to thread without crossing, so too does the whole structure have to be pervaded by a network of supply lines and sewers, and not one of them can go across or below another, there being neither across nor below.

The problem could be avoided by using time. There would be lavatories and filling stations at strategic points where the brain cells could rest and be replenished; but the efficiency of the whole structure would be heavily impaired, and the cells would be able to

The point about sustenance is an allusion back to Chapter 2, where we saw that the mechanisms of life depend upon the driving power of interconnected reactions, and the need to import food. Every step in the function of a brain is a chemical reaction that can occur once when it is triggered. But once it has taken place it has to be rewound by coupling to another reaction having the tendency to run in the right direction. That in turn must be restored once it has occurred, and so one works back to the need to eat, and ultimately to bask in sunshine.[26] On the other hand, a two-dimensional electronic brain, which I think could be prepared only as an artifact of a preceding biological civilization (at least within a reasonable time), could be powered by radiation. Once that has been achieved, there appears to be no particular reason (given that neurons can be made sufficiently compactly in two dimensions) why the two-dimensional human 2-brain should not shrink to around $36\,m^2$ or so and his electronic 2-dog to $15\,m^2$ or so.

Networks that cannot be constructed without intersection are those that contain the structures[66]

  and

These networks are, of course, attainable if the connections between neurons are by radiation (but we shall see problems there soon) because radiation can pursue intersecting paths.

A detailed analysis of the nerve connections involved in the retraction of a gill in the snail *Aplysia* has been referred to already.[37] The neural circuitry for the withdrawal reflex is drawn on p. 67 of the reference in a slightly simplified form. If that complexity is required to withdraw a gill, think what complexity is needed to withdraw an offer.

participate in only a few thoughts before they had to wriggle off between their neighbours, like sheep in a crowded flock, in search of resuscitation.

Bulk also impedes reproduction and retards evolution. In such large creatures evolution would be more like imaginative redevelopment and urban renewal. A particularly languid form of reproduction would have to develop, for heads could barely move. Furthermore, in order to avoid random connections between cells, and to have a chance of producing even one mildly competent genius among populations of vegetables, the network needs to be carefully specified. The genes must carry this detailed specification; yet they too are only two-dimensional and can carry only little information. Not only is ordered complexity unlikely to emerge, it is also liable to decay: the race would be unintelligent and unstable, and intelligence recessive.

If, in spite of all that, a two-dimensional brain did manage to evolve, it would turn out to lack the logical capability of creatures in higher dimensions. This is because not every network of connections in three dimensions can be reconstructed in two without lines doing the impossible and crossing. Even quite simple levels of conscious process in three dimensions, such as the retraction of a gill, appear necessarily to involve networks of connections that cannot be formed in two without numerous unavoidable, and therefore impossible, crossings.

When we turn to brains in three dimensions, the difficulties vanish. Now it is possible to have a richly complex network of interconnections inside a small volume. No longer do heads have to be so unwieldy that they interfere with hunting and reproducing. No longer is there any restriction on the complexity of the logical network that may evolve. No longer do massive power resources need to be applied to pushing in food and pumping out waste. The organization of the sequence of signals is much easier to achieve when transmissions are almost instantaneous. Now the bulk of the brain can be devoted to switches and not wasted on cables.

It might therefore be conjectured that four-dimensional brains would be both super-compact and super-brilliant. That may

The intrinsic differences between the propagations of waves in spaces of odd and of even dimensionalities can be summarized as follows.[22, 67] Single pulse waves (bangs) in odd numbers of dimensions spread radially as sharp pulses, diminishing in amplitude but maintaining their sharply rising fronts and falling tails, giving no warning of their coming and leaving no wake. In even numbers of dimensions, the spreading pulse maintains its sharply rising front, so there is no warning of its coming, but leaves behind a wake as the disturbance continues long after the crest has passed.

I think it worth reiterating the view that while the emergence of inorganic brains (constructed like computers) with their connections either passive conductors (wires or even canals, not nerves) or in the form of radiation is conceivable, they are unlikely to evolve naturally, although being built by a biological system could be regarded as a generalized form of evolution. Complexity has to be gathered in stages when it is not imposed from outside by a designer, and biological systems, almost by definition, are stages in the progressive acquisition of complexity. At an advanced stage of this process, such as we are now entering, it may be possible to discard the unnecessary clutter of biological scaffolding by constructing 'artificially' some desired component such as the brain free of legs, teeth, and intestines just as conventional evolution has shed much of our hair.

indeed be so, but no new logical networks can be constructed when a new dimension is introduced as we go from three to four (or even more). What disarms the brilliance of a four-dimensional brain is its dim perception of its external world.

Most communication is through the agency of waves. There are sound waves, and the whole span of electromagnetic waves responsible for television, communications, and astronomy. A striking feature of waves is that their behaviour is quite different in spaces of even dimension (2, 4,.....) from what it is in spaces of odd dimension (in our own, for instance, with three). In odd-dimensional universes, waves propagate without distortion; in spaces of even dimension, they blur. Whereas we hear 'Bang!' as 'Bang!', because the short, sharp, shock wave travels through the air without spreading, a two-dimensional listener would hear 'Ba..a...n....g...' instead. Although a four-dimensional brain might be super-fast and super-logical, it would perceive the external world only foggily, and it would be insulated from information by the intervening space. The species would be intrinsically short-sighted and hard of hearing, and its members confined to introspection within their own individuality.

Five-dimensional brains could be super-compact, and interact with others and observe the universe with unabated clarity; but they could not evolve. In order to see why this is so, we turn to another aspect of existence.

We take as a hypothesis that in order to come into existence consciousness requires mildly warm, stable platforms that do not experience great variations of climate and conditions, and that persist for long periods. The Earth, if not a paradise, is at least a paradigm.

While planets may be necessary for the emergence of consciousness, they alone are not sufficient. A planet has to be provided with a supply of gentle warmth in order to retain the fluidity of its environment. That fluidity is necessary in order that molecules are endowed with sufficient mobility to explore pathways of evolution. Such warmth could come from within the planet, and it is conceivable that life could evolve on the surface of a warm star. But warmth itself may be insufficient. Life lives off quality (in the sense

Recall the concept of the quality of energy introduced on p. 20 and Dyson's order of merit.[27] Light is a high-quality (low entropy) form of energy; heat is corrupt (high entropy).

The age of the Earth is about 4.6×10^9 years. Life originated about 3.5×19^9 years ago,[1] hence it took over a thousand million years for minimal organisms to emerge. The first multicellular creatures formed about 0.6×10^9 (600 million years ago. Mammals emerged just over 200 million years ago (notice the quickening pace). Man-like ape turned into ape-like man a couple of million years ago, and *Homo sapiens* has been around for at least 100 000 years. Overall we needed almost 5000 million years of gentle baking to come about.

In three dimensions a small disturbance lets the circular trajectory survive if the impact is not too great.[22, 68] In higher dimensional spaces the planet falls on the attracting centre or flies off into infinity. Furthermore, in these higher dimensional spaces there do not exist motions comparable to elliptical motion in three dimensions, and all trajectories have the character of spirals. Ehrenfest points out another entertaining

of Chapter 2). Whereas hot sources could drive substances into the acquisition of complexity, their heat is difficult to control and localize in the places it is required. Heat is simply too dangerous, for it may destroy what it has briefly formed. For the safe, sure-footed evolution of complexity, we need the highly controllable, high quality, localizable form of energy that we find in light.

A planet that was itself hot enough to be a source of light, and hence potentially a source of life, would also be a global crematorium: no complex molecule, let alone corporations of molecules acting as a body, could survive. Of course local sources of light can be engineered on a cool planet, but they normally depend on ingenuity, and hence are a consequence of life rather than its motivation. Except in civilizations, and in hostilities, sustained, persisting, *steady* light has to come from outside a planet; hence planets need suns for life.

If a planet is to be a platform for the emergence of life, it must orbit its central sun without swinging in so close that emerging life-forms are fried, and without swinging so far away that they are frozen. Orbiting must not be erratic if life is to have a chance to emerge, and if delicate molecules are to stumble upon frail and sustained complexity. These gentle conditions, and therefore stable orbits, must persist for aeons. Finding criteria for the evolution of consciousness therefore reduces to finding criteria for the stability of planetary orbits.

I shall now argue that consciousness is three-dimensional. The basis of this view is that only in three dimensions are there planetary orbits of persistent stability, and so only in three dimensions is there time and opportunity for delicate complexity to be assembled to the point where it can respond to its environment with the subtlety that we who experience it term consciousness.

In two dimensions, as in universes of four dimensions and more, planets are easily dislodged from orbits by minor perturbations. The passage of a comet can send a planet down to fry in its sun or off to freeze in outer cold. But the Earth and its analogues near this and other suns, respond to passing comets, other planets, and close encounters of other kinds, and survive. The freedom of motion permitted by the three dimensions of our space

feature of three-dimensional space: the number of rotations is the same as the number of translations (three in each case). In two dimensions there is one rotation (like a gramophone turntable) but two translations (along *x* and along *y*). In four dimensions there are six rotations and four translations.

These other universes have to be distinguished from the other universes that are sometimes invoked in order to provide a framework for explaining quantum mechanics.[69] Such inelegant interpretations (I do not speak of Davies's book) suppose that at each act of observation this universe splinters into slightly different replications. I find it hard to accept that such cosmically profligate ideas can be advanced seriously, but they are.

is just enough, and not too much, to enable them to adjust their paths subtly and to avoid disaster. Disasters avoided, there remains the opportunity for development of selves, and then self-development, until through bloodymindedness, perhaps, the chance is squandered by the petty local impertinence of wars.

Consciousness is a property of minute patches on the warm surfaces of mild planets. On this planet, until this century, it had been confined to the scope of individuals. Here now (and presumably cosmically elsewhere at other times) the patches are merging through the development of communication into a global film of consciousness which may in due course pervade the galaxy and beyond. None of this potential for development of the coordinated motions of atoms into percipience and intelligence could have been realized if the universe had emerged from its creation with other than three dimensions of space.

The argument should really, of course, run the other way. Dimensionality gives the universe the opportunity to develop consciousness; consciousness is not a reason for a particular dimensionality. Man and his counterparts elsewhere are merely elephants with a tendency to hubris. We are fragments of the universe, elephants happily free to roam intellectually as well as spatially. As elaborate outcrops of the physical world, and no more than that, we are no more necessary to its existence than is a breeze. As a universe could exist without a breeze so it could exist without the property of consciousness.

This raises the following question. When a universe stumbles upon its own creation, will it necessarily fall into being with three dimensions? If it does, then it will have the potential to develop the quality of knowing that it has done so (although factors other than its dimensionality might modify, we might say impair, its actual course of evolution and leave our own style of consciousness undiscovered). If it does not necessarily adopt three dimensions, then other elsewheres may be littered with unconscious but nevertheless existing universes. Are we alone, in the sense of all other universes being dead by virtue of their dimensionalities? Or are all the other universes that may exist, pervading other elsewheres and other times, all three-dimensional and potentially inhabited with consciousness?

It may be the case that the universe has more than four dimensions of spacetime, with all but four dimensions rolled up ('compactified') and undetectable. Thus, from a distance, a hosepipe resembles a line, but in fact it consists of a stack of two-dimensional circles; so an actual line in spacetime might be a compactified version of a hypersurface, and our four-dimensional spacetime might be all that is left of a tightly furled spacetime of higher dimension. There is far more than whimsy in this thought, for there is currently a strong indication that we inhabit a ten-dimensional universe, in which six dimensions have been furled up.[25, 57] The radius of the hosepipes that we mistake as lines are about 10^{-35} m, so there is little chance of observing them directly (but every chance of observing their consequences).

Puns on knots are almost unavoidable, even in three dimensions.

A mathematical knot can be pictured as a knotted piece of string with its ends spliced so that it cannot be untied.[70, 71] There exist two basic knots in three dimensions, and they cannot be deformed into each other:

Two-component structures are of extraordinary importance in quantum theory.

The theory of the stability of knots is related to the study of *solitons*,[72] which also provide a description of the stability of things, such as particles,[73] in terms of the topology of space.

The currently most favoured fundamental theory of spacetime is *string theory*, in which the basic elements of the space are not points but linear entities (hence 'strings').[25, 57]

That other intelligences exist in this universe is now reasonably certain, for our understanding of the formation of stars suggests that accompanying planets are a common feature. That other intelligences do not exist is so improbable as to be unworthy of further qualitative speculation. Our concern here is the same problem cosmically magnified: do we know enough about the formation of *universes* to be sure that other modes of intelligence exist outside this universe? Are we hypercosmically alone?

On what grounds, then, should a universe adopt three dimensions or any other number when it comes into existence? We are disregarding reasons based on purpose, regarding them as superfluous. Therefore, the reason must be found in its fitness to survive. What is it that survives in three dimensions, we have to ask, but not in others? We have to seek reasons why the universe can survive its own creation, and we must relate those reasons to its dimensionality.

A knot is a simple example of something that can survive in three dimensions but not in any more. You can't even tie a knot in two dimensions, because you cannot cross over or below (but that does not rule out the possibility of two-dimensional things wearing woolly vests). You can make involved ravellings in four dimensions, but they are not knots and just pull apart. In three dimensions you can knot, in contrast to two where you cannot. The knots can persist in three dimensions but they cannot persist in four.

The persistence of knots in three dimensions so closely resembles the persistence of particles, that it suggests that particles are nothing other than knots in spacetime. Different species of particles are like different species of ravellings of spacetime, and they persist in time because three dimensions of space eliminates the possibility of them simply unravelling. The stabilities and identities of particles then boils down to dimensionality: three dimensions of space is the minimum for existence and the maximum for persistence. In universes of other dimensions there would not merely be no property of consciousness, which is an enjoyable irrelevance, but there would not even be matter.

We can, I think, now begin to see the source of forces. A knot of spacetime, a particle, is embedded in spacetime, and its twist

This is pure speculation, and is too vague to be science, but I think the idea of knots embedded in space and actually distorting it or, what may be equivalent, generating other knots locally which then propagate until they are in the vicinity of another similar knot, which then acts as an antenna and responds to the passing knot, does have such striking similarities with the known forces and their interactions as to be suggestive of truth. The picture is akin to the topological explanation of the electromagnetic interaction.[11, 44] In that explanation space is regarded as being full of wormholes, with an entry in one place and an exit elsewhere. Electromagnetic lines of force then pass through the hole. An observer sees them disappearing at one point (and calls it a charge of one sign) and appearing elsewhere (and calls that an opposite charge). Other types of force could then be regarded as other topological distortions of spacetime.

The equations of general relativity in spacetimes of various dimensionalities have been considered.[74] In two-dimensional spacetime we can have curvature but not matter (implying arbitrary curvature), and in three-dimensional spacetime space must be flat, implying that matter cannot interact gravitationally. The exploration of the field equations in spacetimes of high dimension, the *Kaluza-Klein theories* mentioned above, is currently the target of considerable interest and some success in the unification of gravitation with the other forces of nature.[25, 57]

The *Maxwell equations* describe the electromagnetic field. They were the first step in the unification of the forces in the sense that they show that electricity and magnetism are intrinsically the same. The uniqueness of the Maxwell equations in terms of the dimensionality of spacetime has been investigated.[75] A number of observations indicate how four dimensions are preferred. For instance, the Maxwell equations have a symmetry which is revealed only in four-dimensional spacetime, and only in four dimensions can one perform a particular technical operation on the electromagnetic field which can also be applied to the gravitational field. The investigation builds on a remark made by Einstein that 'the gravitational equations for empty space determine their field just as strongly as do Maxwell's'. It is shown that

twists its surroundings locally, and that twist in turn twists, and the warp is propagated into the distance. Gravitation, or the warping we call gravitation, is then just the consequence of the embedding into spacetime of its own knots. The greater the mass of the object, the greater its knottiness, and so the greater its distant influence. Knots of particular kinds induce different kinds of knottiness in the space nearby, and these few knots propagate and interact with like knots elsewhere. Forces are beginning to emerge.

Yet the existence of particles, and the persistence of identities, is no more a reason for three dimensions than the existence of elephants and intelligence. Unless we can find another reason for the emergence of three dimensions we shall have to admit the possibility of amorphous universes. So far only prejudice denies their existence. We have to dig deeper.

Gravitation is known to have properties that reflect the number of dimensions. For instance, if spacetime had only one dimension of space (together with one of time) there could be no matter, and space could adopt curvatures of random magnitudes. On the other hand, if there were two dimensions of space, matter could exist but the intervening space would necessarily be flat, and so objects could not interact. Only in three-dimensional space, space like ours, can matter both exist and propagate its influence to both its immediate and its distant neighbours. The existence of three dimensions permits the members of the universe to be coordinated into an entity. In less than three dimensions they would either take no form, or forever remain totally isolated.

In a universe of intrinsic simplicity, but rich in properties, there needs to be a variety of forces. A force as universal as gravitation, however, then has to be compatible not only with a force as universal as electromagnetism but also with all the other types of force that bind things into elephants. In any other dimensionality of spacetime other than our own, these forces appear to be mutually incompatible. Only in a spacetime of our dimensionality, three dimensions of space and one of time, are forces compatible with the existence of matter.

Time's dimensionality is also not an accident. If time had more than one dimension you could turn round in time just as you can

only in four dimensions does this remark apply. The significance of this result appears to be that electromagnetism and gravitation can occur jointly in a unified field theory only in a four-dimensional spacetime that has undergone compactification to four dimensions.

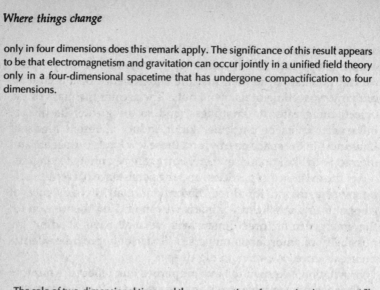

The role of two-dimensional time and the conservation of energy has been noted.[76]

In the case of $(+,+,+,-)$ the radius of the universe can begin from zero and reach its maximum value at some later epoch, then collapse. In the case of $(+,+,+,+)$ the universe moves towards a minimum non-zero radius or away from it. There is no time at which the radius of the universe is zero, and so there is no 'big bang'.[77]

turn round in space. But as time has only one, you are left to pursue the future. The structure of our spacetime ensures that consequences of present actions lie in the future; that would not be so if time were dimensionally richer. Newspapers with two dates would not necessarily report the past.

The death of causality, and the preceding of events by their causes giving way to disconnected causes and events, would have a stronger impact than confusion. It would entail the death of being. As I shall go on to explain, newspapers with two dates, like their readers and their readers' atoms, could not endure. Having too much time, in the sense of having several dimensions, is as dangerous to duration as having too little.

Too much time allows too much freedom to circumvent a special type of obstacle to change. In particular it allows freedom to circumvent the constraints imposed by the conservation of energy. This has a chain of disagreeable consequences. The conservation of energy implies the conservation of matter; therefore if energy is not conserved, then nor is matter. Moreover, the conservation of matter is the conservation of the creases and knots of spacetime. Therefore with the elimination of the conservation of energy goes the elimination of the structure of spacetime. Spacetime is stabilized against collapse by the conservation of energy; but with another dimension of time this constraint crumbles, and with its crumbling crumbles matter. The universe would be over in a flash if time had more than one dimension. Overendowed with time, it could not survive its own creation.

But why is there time at all? Time is not here for the benefit of our comprehension; we merely take advantage of it drifting, perceiving, unavoidably into the future. What brought about time? Why does the universe come into existence with a time, and not merely as space hanging in existence outside time, being free of eternity?

An answer is beginning to emerge. It runs as follows. If spacetime lacked time and were truly four-dimensional space, with $-(ct)^2$ in its geometry replaced by $+(ct)^2$, then progress in either direction along this fourth dimension would never bring the universe to a point. With our type of time, the whole universe can be traced to a single point, and had a beginning. With that alterna-

tive space-like time, the history of the universe would never have had an identifiable beginning. Without time, in our sense, there is no beginning. With a beginning, in the sense of there being at some point in spacetime a pointlike creation, there is necessarily a geometry with the complexion of time.

Where are we? We have seen that the arena for all events is spacetime, space and time amalgamated yet distinguished by the adoption of a special geometry. The local distortions of spacetime, its knots, are the fundamental particles and forces of the universe, and they persist on account of the dimensionality of spacetime and the peculiarity of its geometry. That special geometry permits stability and inhibits the immediate evaporation of a newly formed universe. We have seen that the existence of time is a necessary concomitant of creation, and as a bonus it brings with it the distinctions of causality. Causality renders the universe potentially comprehensible, and allied with the properties of three dimensions of space, brings it to the stage of being actually comprehended.

The characteristics of spacetime, the very ones that make it structurally stable, have also happened to permit its evolution to the stage of having outcrops of consciousness. That consciousness is now embedded in us and in others, and is rich enough to be capable of elaborating simplicities into art and simplifying complexities into science.

Fourth Orientation

We have seen that matter and energy are spacetime, and that the dimensionality of the universe is such as to allow matter not only to exist but also to persist. The slow and interconnected unwinding of the initial creation—what we have argued should be regarded as natural, spontaneous, and purposeless collapse into dispersed and chaotic uniformity—brings forth consciousness and crops, purposes and pots, motives and machines, and belief and understanding as ephemeral efflorescences. That is what we, who are some of the perceptive outcrops of the universe, experience as being. We have seen that in the creation there needed to be formed space and time; then with our familiar dimensionality it is able to display the richness of properties we perceive as matter, energy, and forces, and to be the arena for events, events of such subtlety that in places things become self-aware. In a deep sense, spacetime itself is self-aware.

One form that this self-awareness takes is in the ability of brains to generate mathematics, and it is mathematics that we use as a language to discover and express the new. That mathematics is the profound language of nature, the apotheosis of abstraction and the archenabler of the applied, should stir us into wondering whether herein lies the biggest hint of all about our origin and the origin of our understanding.

Five

CALCULATING THINGS

ONE OF THE DEEPEST PROBLEMS OF NATURE IS THE SUCCESS OF mathematics as a language for describing and discovering features of physical reality. In short, why does mathematics work? Why is it that Newton's equations can be used to predict, at least approximately, the actual behaviour of a particle? Why is it that Euclidean geometry is successful in local regions? Why does quantum theory appear to be an exact description of the world, even though we might not fully understand it? Why do predictions based on Einstein's field equations turn out to be actually observable? Why, as we grope ever deeper into the past in search of our origin, is mathematics the ineluctable handmaiden of experiment? Such questions are of extraordinary interest and importance, and it may be that the answer to them will reveal all that there is of the soul of the universe.

We humans have stripped back the clouds that cloak our understanding of our cosmic beginning and our current persistence to the stage that exposes the mathematical structure of the world more clearly than it has ever been observed before. (Other beings elsewhere may already have stripped back understanding to the bone, but of that there is no evidence and we must act as though we were alone.) At our current level of study of the substructure of the world we may be examining the bedrock of being. By stripping elephants down to elements, and elements down into quarks, we are unpacking the complexity of the universe and exposing its simple heart. The process is revealing what may be the deep structure of the universe, and—for the sake of this current discourse—showing most clearly the relation of that structure to mathematics.

Furthermore, the attention of seriously equipped thinkers, those thinkers we call scientists, is at last beginning to turn to that other great conundrum of being: consciousness. That we may finally be capable of progress in the resolution of this conundrum, even though so little progress has yet been made, is relevant to our exploration because our brains appear to be the springs of mathematics. If we can understand why that supreme construct of the human intellect, that archdisembodiment of intellect, mathematics, works as a description of the world, then maybe we shall

The reference is to the introduction to *The legacy of Greece*,[78] to which Murray contributed.

The *tetraktys* is

```
   *
  **
 ***
****
```

The Pythagoreans were amused to note that it generated the first four integers. They were thrilled to note that those integers were the numbers of points needed to construct objects of successively higher dimension

have an insight into cognition. Almost literally, by looking deeply into the universe, we may see inside our own heads.

It is possibly not too extravagant to claim that the answer to the question of why mathematics works will be the final answer to all questions of being (including those questions that are driven by sentimental wishful thinking). But we must not fall into the accommodating trap that ensnares sedentary philosophers: we need experimental science to guide us through the labyrinth of complexity that is so characteristic of the external world. Experiments focus our powers of producing mathematical descriptions of tiny fragments of the whole, fragments that can be examined to test whether we are right. But once we have arrived at the appropriate mathematics, we shall have an extraordinary understanding, perhaps the deepest possible understanding, of what it—our cosmos—is all about.

The Greeks, of course, thought much the same. Gilbert Murray once remarked that Greek thought was marked by a clarity of vision that may be typical of those founding a new discipline:

'they started clean from nature, with almost no entanglements of elaborate creeds and customs and traditions'.

Unfortunately, their primordial flash of inspiration and keen insight, if that is what it was, was left stranded on the beach of armchair argument by the absence of equally penetrating experimental support. In the absence of that critical reinforcement, the enemies of light invaded and two thousand years of obfuscating decoration were piled on what may have been glimmerings of the truth. Only now are our experimental abilities becoming a match for the power of our mathematics, and it may be time to return to the original Greek vision.

The Pythagoreans, for instance, may have attained absolute insight unknowingly in the recognition of the *tetraktys*, and maybe there is a mechanism beneath their discovery that we have yet to discover. But that was where their approach effectively stopped. They were unable to develop it, maybe because it was wrongheaded and empty, but quite possibly because simple concepts like relations between numbers are largely irrelevant to everyday objects, almost the only things they knew quantitatively. They saw

```
*        1, a point
**       2, a line
***      3, a triangle
****     4, a tetrahedron
```

But they were rocked on their intellectual heels when they saw that the ratios of successive lines were the principal harmonic intervals:

2:1   octave
3:2   fifth
4:3   fourth

They sensed that harmony and hence beauty were allied with the structure of space.

what is maybe a glimmer of truth where they were confronted with a simple abstraction, harmony, and with an uncluttered arena, spacetime. But then the dark night of ignorance, and then the flood of misdirected scholarship, enveloped the glimmer. Now, however, we are starting to deal with what may be the truly basic, exceedingly simple, objects, and it may be that the ideas of the Pythagoreans, having slumbered and been scorned in their slumber for centuries, will now come into their own.

Plato, for instance, supposed that mathematical objects were really there, and were not mere creations of the mind. It may be that, in a certain sense we need to establish, Plato was right, and there is a certain reality in a mathematical object—a reality about the integers and an actually existing perfect circle. But is it also possible that his views need to be interpreted in a more modern idiom. Whether or not we judge that Plato proves to be 'right' because after interpretation his views can be substantiated is of wider interest because it has implications for our attitude towards mythical accounts of the creation. Should we regard the Genesis account of the creations as 'right' because we can see resonances with what we think actually happened, even though the Genesis committee of authors did not really know what they were talking about? All religions, except science, acquire authority from ambiguity.

Heraclitus of Ephesus took the view that the reason why we can comprehend the universe is that each of us contains the *logos*, the divine spark. A modern Heraclitus might take the view that the same divine spark can generate the mathematics that we use to describe and discover the world, so it is hardly surprising that our cognition of the world can be expressed and interpreted in terms of mathematics. Heraclitus did not really know what he meant, and was merely producing a framework for his thoughts rather than the thoughts themselves; I shall do the same in the embers of this essay.

Pythagoras, Plato, and Heraclitus all had the right spirit, but it is probably we, not they, who can begin to see what they may have tried to mean. He who disinterred our current question from the clutter of the centuries that followed its first formulation was

For a history of Einstein's toying with the cosmological constant (the *ad hoc* addition to his field equations that prevents the expansion of the universe), see the scientific biography of Einstein by Pais.[79]

The uncertainy principle states that the position and momentum (the product of mass and velocity) of a particle cannot be specified simultaneously with arbitrary

Copernicus. The Copernican revolution was only superficially concerned with whether the Sun circled the Earth, or vice versa. Its deep concern was with whether mathematical models are descriptions of reality. It is possible to predict the motion of the planets in either the geocentric Ptolemaic or the heliocentric Copernican systems, and in certain cases the former may be easier to use because it centres the origin of the coordinates on the point of observation. The Copernican revolution, however, invites us to select the Copernican system as being the 'true' picture of reality. The sleeping thought has awoken that our mathematical formulas are models of the true world.

If we leap forward to our present era, then we cannot fail to be impressed by the success with which mathematics does seem to account for the world. Quantum mechanics can be used to calculate properties of the electron accurate to about ten significant figures. Einstein's field equations are not merely elegantly simple, but have consequences that turn out to be experimentally verifiable—in short, true. Einstein did not try to predict the expansion of the universe; indeed, he found that his equations predicted it, hated the thought, and cobbled together an uglier theory to prevent it, only to be taken aback when the expansion was found actually to occur. Now we use mathematics to construct models of the fundamental particles and their interactions and are coming excitingly close to making reliable predictions about their consequences. Why can this be so? Why does mathematics work? Indeed, why does it work so well?

It should never be forgotten that mathematics and observation jointly squirm towards the truth. The process of discovery of the world is often a sequence of alternations between observations and mathematics in which the observations are stretched like a skin on to a kind of mathematical template. We refine and bootstrap ourselves into a mapping of the physical world by squirming forward, constantly comparing our expectations based on our current theory with observations they themselves suggest.

One example of the procedure is provided by quantum theory, and in particular the uncertainty principle. Although most people appear to consider that the uncertainty principle abolishes any

precision: the greater the precision with which one property is specified, the less the precision allowed to the other.[18, 80] Heisenberg devised a number of thought experiments to show that all conceivable measurements were consistent with his principle. Some have gone on to interpret the principle as limiting our powers of measurement: I prefer to think of it as limiting the richness with which a state may be specified. Thus, we must choose to specify it in terms of positions *or* in terms of momenta. Classical physics falsely presumed (in the absence of contrary evidence) that both descriptions could be applied simultaneously. All (I think all) the conundrums of quantum measurement stem from our determination to abide by our classical conditioning, and to insist (usually covertly) on thinking in terms of positions and momenta simultaneously; the conundrums disappear with the imposition of a rigid self discipline. The uncertainty principle is a great simplification in the sense that it instructs us to focus on one or the other sheets of description. Not everyone has come to terms with the uncertainty principle in this way, and some profound thinkers continue to argue that it is profoundly enigmatic. What is clear (in my view) is that the uncertainty principle does not limit our knowledge of the world: it readjusts our classically conditioned view of what is knowable.

Popular accounts of speculations about theories of everything (TOE) currently sell like hot cakes on doomsday; for an excellent survey, see Barrow's book.[81]

chance we once might have thought we had to comprehend the world, it is more optimistic (and perhaps more correct) to consider that the uncertainty principle is an indication that our classically inspired template for understanding the world is over-elaborate. Classical physics, the physics of the farmyard of everyday experience, forged a template that lead us to expect that we should be able to describe the world using the language of speed and location simultaneously. Quantum mechanics takes its axe to this naive, superficial view. It reminds us of what should be obvious: that farmyard-inspired theories may be too gross and unsophisticated, too covered in the dung of their own origin. It provides a template that in effect requires us to choose one language or another. It tells us to speak in terms *either* of location *or* of speed, and never to mix the two. It tells us to speak German for complete sentences or to speak English for complete sentences. It warns us not to start a sentence in German and then end it in English. Quantum mechanics tells us that the mathematization of Nature should be done using formulas drawn from the language of position or from the language of speed. It instructs us to separate the muddled classical template into two sheets and to use either one sheet or the other. Quantum mechanics clarifies our vision of the world and in so doing exposes more sharply its mathematical structure. That is just one example. In the end, if there is an end, we shall possess a mathematical theory of the universe that matches it in every test: the fit of reality to the template will be exact and we shall have a theory of everything.

To make any progress with our question we should explore some of the hints that seem to be littered around. Hints are very good things for they may be rainbows that mark buried intellectual gold. In what pages remain, we shall seek to identify hints, allow them to effloresce extremely, and observe whether anything worthwhile pops out. Sometimes we shall run up against a brick wall. That must be so, for no one yet knows why mathematics works. But once a wall has been located, someone else might find a way around it.

Ours is such a difficult question that it may be appropriate to begin by setting out a sketch of the answer to use as a framework

For an introduction to Chomsky's ideas, see the short book by John Lyons.[82] This book contains an excellent annotated biography.

for the rest of these remarks. We shall ignite a speculation; since it will carry an air of authority with it if it also bears a name, it will henceforth be called *the hypothesis of deep structuralism*. The name deep structuralism is intended to convey the idea that the physical world has the same logical structure as mathematics. By implication, the reason why mathematics works as a description of physical reality is that they share the same logical structure.

It will also be appropriate to distinguish between the weak and strong forms of deep structuralism. By *weak deep structuralism* I shall mean that mathematics and physical reality merely share the same logical structure and mathematics is a mirror that can be held up to nature. By *strong deep structuralism* I shall mean that mathematics and physical reality do not merely share the same logical structure but are actually the same. In other words, according to the hypothesis of strong deep structuralism, physical reality is mathematics and mathematics is physical reality.

Before we see whether there is any justification for this hypothesis in either of its forms, it may be helpful to say a few words about the origin of its name. It springs, of course, from Chomskyan linguistics, where it is supposed that brains have an innate, hardwarelike capacity for language. Speech is then a kind of surface structure that is generated—transformed—from an innate structure. All human brains seem to have the same kind of innate structure, for the actual language that a child learns appears to depend only on its environment: a Japanese child becomes a native speaker of Japanese if it is brought up in a Japanese milieu, but it becomes a native speaker of English if it is brought up in an English milieu instead. The hardware is innate and essentially identical but the surface structures are different. A typical diagram showing this relation is

Deep structure $\xrightarrow{\text{Transformation rules}}$ superficial structure

Language $\xrightarrow{\text{Japanese, English, . . .}}$ Speech

This relation can be expressed with a somewhat more technological slant:

Hardware $\xrightarrow{\text{Software or firmware}}$ Output

These mappings will reappear later when we consider the acquisition of abilities, a talent that will emerge as a key feature in our comprehension of the creation.

The principal point at this stage of the discussion is that the hypothesis of deep structuralism supposes that two more mappings can be added to this table:

$$\text{General logic} \xrightarrow{\textit{Specific procedures}} \text{Mathematics}$$
$$\text{General logic} \xrightarrow{\textit{Physical laws}} \text{Physical reality}$$

If you prefer weak deep structuralism, vanilla structuralism, then you would regard the last two lines as mutual analogies. However, if you prefer strong deep structuralism, strawberry structuralism, then you regard them as identical. So that we can have the whole framework bare before us, we shall add one more line to the table:

$$\text{General logic} \xrightarrow{\textit{Neural networks}} \text{Consciousness}$$

These ideas will be elaborated and illuminated as we proceed, but it is already possible to see that even the vanilla speculation accounts crudely for the ability of mathematics to describe the physical world, because there is in some sense a resonance between the structure of mathematics and what will turn out to be the fundamental structure of the physical world. The reason why we may be conscious of the world, including the inner, introspective world of emotion and intellect, may be that our brains are material portrayals of the same deep structure. That may also be the reason why brains can generate the mathematics that we need to comprehend the world. There may be a spark of truth in the concept of the *logos*, but in place of shared divinity there may be shared structure. If that is so, then there is hope for us in our quest to comprehend everything—literally everything, for our brains are equipped to deal with whatever there is.

The mathematical description of the world depends on assembling numbers in a certain way. The formulas of theoretical physics, for example, are generalized statements about quantities. Thus, Newton's equation of motion, *Force* = *mass* × *acceleration*, is a generalized statement about numerically expressed observations relating to motion and forces. It summarizes a whole class of

Now, although I may appear to be suggesting that the way forward, the last sprint to the post of total comprehension, may be the resurgence of a kind of metaphysics, I must emphasize that I see it as essential that we test our ideas experimentally at every stage. It is absolutely essential that we test any metaphysical speculation against the results of experiments. That has been the mainstay of science's success for three hundred years

behaviour irrespective of the identity of the object undergoing acceleration and the actual numerical values that are selected. A formula in theoretical physics is a *generalized* summary of relations between measurements. The numbers that such a formula summarizes are also generalizations. Specifically, numbers are generalizations of lists. Thus, the number 2 is a generalization of 'Smith and Jones' and 'apple and apple'; numbers arise from sets, and arithmetic is an aspect of the logic of sets.

The relevance of these remarks is that numbers, as well as summarizing the essence of sets, are also the universal currency of calculation and exposition: physics did not truly begin until people started to express observations numerically and seek relations between numbers stemming from different classes of observation (for instance, those relating to force, mass, and acceleration). The manipulations of numbers that are characteristic of physics are really statements about logic, and writing $6 = 3 \times 2$, which is a very particular case of Newton's equation in a certain choice of units, is actually a condensed version of a statement in general logic.

We may see, even though in our hearts it may be difficult to accept, for we have a fierce sentiment toward retaining the tangibility of the tangible, that there is a kinship between observation (our only contact with reality) and logic. Aspects of the universe are summarized by mathematical formulas; formulas are generalized statements about the relations between quantities; those quantities are expressed numerically; hence, formulas are statements about the relations between numbers; statements about numbers are in fact statements in general logic. Thus we discern the hint that the formulas of physics are expressions about some underlying logical structure of the universe, which is the content of deep structuralism.

Now, it would be perfectly reasonable to stop at this point and to say that there is no possibility of distinguishing between the flavours of deep structuralism. However, there are hints around us that point to the validity of vigorous strong deep structuralism as distinct from pusillanimous weak deep structuralism, and hence that the physical world *is* mathematics.

The biggest hint of all is that we seem to exist, and therefore the

and it would be foolish to discard its procedure now. However, inside our timidity we must not be small-minded: if we have confidence in our science, we should be prepared to launch out on a different mode of understanding. Conventional science has guided us to the kind of mathematics that works: let us cautiously seek to read the message within its success.

An account of the emergence of numbers from absolutely nothing in a richer (but still fundamentally empty) way has been given by the ever-fecund mathematician John Conway (among whose other credits is the invention of the game of life).[83]

mathematical description of the origin of the world must accommodate the emergence of the apparently something from the certainly nothing. We have seen that the seemingly something is actually elegantly reorganized nothing, and that the net content of the universe is now the same as it has always been, and always will be, world with or without end: namely, nothing. What mathematics can emulate the emergence of something from nothing? Why, nothing other than the pure numbers!

Thus, to see numbers emerge from the empty set $\emptyset$—the set with null content: absolutely nothing—we first name it '0' and write $0 = \emptyset$. The set that contains the empty set is named '1', so we write $1 = \{\emptyset\}$. By the number 2 we mean the set that contains both the empty set and the set that contains the empty set: $2 = \{\emptyset,\{\emptyset\}\}$; by 3 words almost fail, but it should be clear that 3 is the name of $\{\emptyset,\{\emptyset\},\{\emptyset,\{\emptyset\}\}\}$, and so on through 4, 5, and indefinitely and ever more labyrinthinely after. We have already seen that the numbers are the currency of physics, and here we have them popping out of nothing, the empty set.

We have previously argued that there is nothing in the universe at all. Now we should note that the numbers are still essentially collections of the empty set, they are still assemblages of absolutely nothing. Although we apparently have a collection of numbers, we still truly have merely sets of sets of the empty set. We have an illusion of something, not something itself, the origin of mathematics is a mirror, at least, of the origin of the cosmos. It may be that the logical structures of mathematics and the universe emerged simultaneously and are identical. The deep structure of the universe may be a globally self-consistent assemblage of the empty set. We, like mathematics, and like it or not, are elegant, self-consistent, reorganizations of nothing.

The kind of question that might now arise is whether *everything* mathematical has its counterpart in the real world. If the physical world is mathematics, then should not every mathematical structure exist in it? That can be expressed slightly differently: is all mathematics waiting to be discovered in the physical world? If only weak deep structuralism is true, then this is not necessarily so, for there may be only a homomorphism rather than an isomorphism

between reality and mathematics. However, if strong deep structuralism is valid, then presumably reality holds a physical entity that is the counterpart of each and every mathematical structure.

At first sight, that would appear not to be case. We can write down the Pythagoras theorem for a space of 1000 spatial dimensions,:

$$(\text{hypotenuse})^2 = x_1^2 + x_2^2 + \ldots + x_{1000}^2$$

but as we do not appear to live in a space of 1000 spatial dimensions, this expression does not seem to have a counterpart in reality. Likewise, we can lay out the mathematics of the Bohr theory of the hydrogen atom, but that model is known to be false. So at first sight it looks as though strong deep structuralism cannot work.

There may be two ways out of this difficulty. One is defeat, always the easier option for the faint-hearted and sometimes the wise. We could resort to weak deep structuralism, assume that mathematics can throw up a froth of many classes of object, and then accept that only some of those objects have their physical counterparts. That is weak-weak deep structuralism. Suppose, though, the hints that have led to the strawberry speculation are so compelling (or the desired end so magnetic) that we have confidence (or at least are possessed with foolhardy determination) that that idea can be salvaged. To do so, we may have to distinguish between the universe, which must presumably be self-consistent globally, and a local entity, a brain, which can generate mathematical structures free of the constraint that they need to be consistent with the structure of every electron and the motion of every planet. The mathematical structure we call the universe may have to be simultaneously, globally, and perhaps nonlocally self-consistent. *Our* mathematics, the statements we make on paper, need in some sense be only locally self-consistent. Thus, instead of seeing that the 1000-dimensional Pythagorean theorem eliminates strong deep structuralism, its notional, local existence may hint that the theory of the universe should be *nonlocal* strong deep structuralism.

These speculations lead us naturally into the question of brains

and their functions. It will be remembered that a part of the speculative framework of deep structuralism is that brains can generate mathematics because in a sense they are chemical (biologists would say biological and physicists physical) portrayals of an underlying logical structure: brains are material realizations of logic. That we can be conscious of the world is, at its deepest level, a manifestation of the resonance of the deep structure of a brain with the deep structure of the universe.

There is a dangerous door open here, saying 'Eat me'. We could Alice through the 'Eat-me' door and achieve false enlightenment by adopting the view that, because brains are built from the same stuff as mathematics, then they are naturally disposed to understand it and create it. This approach is far too facile. I suspect that there is a deeper reason for the ability of brains to resonate with and commentate upon reality. Buttercups do not think, yet they are also built of mathematics. If buttercups do not cogitate, but we do, yet are built of the same ultimate stuff, then the difference must lie in the complexity of our structures that has emerged from the process of evolution.

Brains that are capable of generating the mathematics that reflects reality evolve as a by-product of learning to cope with the environment. Mathematically capable human brains did not evolve because there were selective advantages in being able to solve quadratic equations or to write tensor field equations. There is no need to be able to solve Newton's equations, let alone Einstein's, when you are a monkey: it is better just to jump out of the way. Yet somehow the skills needed for survival developed exactly that faculty; and by virtue of their nonlinearity, brains appear to have transcended evolutionary requirements.

This is a hint, perhaps, that the physical structure of the brain is a material portrayal of the logic that the being encounters in the world. To some degree, the capacity is inherited, but a component of its environmental. Once again, an analogy from linguistics may provide the hint we need. A child is capable of becoming a native speaker of any language; an adult, in general, is not. At least some types of learning are correlated neurophysiologically with the loss of synapses. If the theory of the maturation of neural networks by

I must confess that I feel rather bewildered at this point. No one yet knows what consciousness is, but thinking of it in nonlocal strong deep structuralist terms in the same way as we might think of the union of mathematics and physical reality may turn out to be the basis of a helpful, global, systems approach to our comprehension of comprehension.

the selective stabilization of some synapses and the elimination of the surplus is correct, then we can think of the learning of Japanese as a kind of withering away of the structure corresponding to the transformation rules that are not needed for Japanese, and the process of learning English as the withering away of the synaptic structure corresponding to the transformation rules that are not needed to generate English as the superficial structure. This eliminative approach to learning would appear to be consistent with the difficulty that native speakers of one language have in learning another later. It also neatly accounts for the fact that different individuals have neural networks with widely different morphologies, yet they end up thinking similar thoughts and speaking the same language. The brain has evolved as a highly redundant network, a generalized, semi-soft computer, that can have aspects of its environment impressed on an exuberantly sprouted and only partially genetically predetermined network.

Somewhere here it may be dimly perceived that a brain develops as a network that is, like the universe itself, globally self-consistent. It is fitted for this development by evolutionary pressures that lead it to have a plastic morphology able to accommodate and process perceptions of the world. In the course of developing its processing abilities, it unconsciously discards those synapses that render it less than globally self-consistent, leaving it a microcosm of the macrocosm. That is, the deep structure of the brain may be in resonance with the deep structure not only of mathematics but also of the physical universe. But here I run up against a wall.

However, let us pass on from here and look as some of the implications of deep structuralism. A positive point is that we can currently perceive the emergence of a new pardigm of science, one that may be heaven sent, so to speak, to deal with the deep questions we have been exploring. I suspect that the days of our analytical approach to physics may be numbered. The major thrust of mathematical physics until recently has been the formulation of differential equations and the discovery of their analytic solutions: the wonder of physics has been (and still is) the power with which an analytical expression summarizes a great swathe of observation

Some of the brain-invading images that may be computed, and which have given birth to a new paradigm of art as well as science, have been published[84, 85] and are indeed available in a form to frame and hang. The mathematics of the images has been explored by Schroeder[86]. The difference between a fractal image and a Mona Lisa is that when the latter is inspected closely, one sees an ever more amorphous blob of paint; however, when a fractal image is inspected, one sees ever more detail. Fractal art is art without end. Viewing it is like standing on a precipice of enthralling beauty over a chasm of infinite depth.

and prediction. The new paradigm of explanation that seems to be emerging is based on the view that even simple equations might have consequences of profound complexity. Thus, although we may know the seed of a pattern and can specify the very simple rules of transformation of the seed, we cannot predict the outcome of the repeated application of those rules of the initial seed. Extraordinary complexity can burgeon from this intrinsic simplicity, as can be appreciated by examining the mind stirring images that can stem from fractals, the Mandelbrot set, and cellular automata. Such images are of such endless beauty that in them we have transcended the tetraktys and its harmonies and have touched perhaps the springs of our perception of beauty. Mathematical physics and aesthetics have merged, and science is ready to become whole.

It is already possible to see where such concepts can be applied to the description of the real world. We know the seed in the gene—the DNA molecule—and we know the rules for the transcription of the base sequence into proteins; but maybe, like a pattern of infinite complexity, there is no chance of ever predicting the morphology of a being even if we knew its DNA. The same may be true of signal processing in the brain. Although we may in due course know all the circuitry of an individual brain, and particularly a simulation of a brain, and can model its multichannel modes of neurotransmission, we may never be able to make a prediction about the outcome of a given input except by simulating it directly.

Not only has the theory chaos endowed us with a new paradigm of understanding, it may be *exactly* what we need for the mathematical description of consciousness. The fact that we need a computer to explore the ramifications of complexity, and a binary computer is the simplest conceivable type of computer, may point to a deep analogy between Boolean logic and the workings of our brains. Binary computers may mirror *precisely* the deep structure of the universe.

Fifth orientation

It may be that the comprehensible is comprehensible because of the deep structural resemblance, and even perhaps the actual identity, of mathematics and reality. The Greeks may have been right. The origin of numbers from pure emptiness may prove to be deeper than a mere analogy between the emergence of a system and the emergence of a cosmos: perhaps in the deployment of emptiness we shall be able to discern the future form of a scientific theory of creation.

The coming into being of space and time is the central event of the creation. So now we turn to the problem of expressing the processes that brought it into being. First we shall think of something that could be ordered into what we now recognize as spacetime. Then we shall have a glimpse of how it may be possible to allow the helping hand of the infinitely lazy creator let slip the last fingernail of assistance. The necessity of the creator will be seen to fade. Then as the creator drops out of involvement, so the universe comes into being without external intervention, out of absolutely nothing.

Six

CREATING
THINGS

The Big Bang has been the subject of some criticism in recent years, and some people have been misled into believing that it has failed. In fact, the standard relativistic hot Big Bang model remains the best theory we have of the origin of the cosmos; there are no well-established contradictions with observation, and the number of predictions and interpretations that it has been used to make substantially outnumber the elements used in the formulation of the theory. The case in favour of the theory is argued by Peebles, Schramm, Turner, and Kron.[87] Whether the more elaborate versions of cosmogenesis are viable, such as the inflationary universe,[25, 57, 88] in which it is supposed that the current era of expansion was preceded by a brief instant of colossal expansion, is still very much an open question. Very recent (April 1992) observations of texture in the microwave background add impressive experimental supports to the Big Bang scenario and guardedly optimistic support for an inflationary era.

The idea of a pre-geometry, a dust of unstructured points, appears to be due to Wheeler.[11] For 'dust of structureless points', read 'a Borel set of points not yet assembled into a manifold of any particular dimensionality'. Throughout this chapter and the next, it might be helpful (and if not helpful, thought provoking), to bear in mind the description of the emergence of numbers from the empty set, as outlined in Chapter 5.

The modern discussion of phase transitions is in terms of the *renormalization group*, which can treat various classes of system in any number of dimensions.[89]

NOW WE GO BACK IN TIME BEYOND THE MOMENT OF CREATION, to when there was no time, and to where there was no space. From this nothing there came spacetime, and with spacetime there came things. In due course there came consciousness too, and the universe, initially nonexistent, grew aware.

Now, at the time before time, there is only extreme simplicity. There is really nothing; but to comprehend the nature of this nothing the mind needs some kind of crutch. That means we have to think, for the moment at least, about something. So, just for the moment, we shall think of *almost* nothing.

We shall attempt to think not of· spacetime itself, but of spacetime before it became spacetime. Although I cannot explain exactly what this means, I shall try to indicate how you can begin to envisage it. The important point to appreciate is that it is possible to conceive of structureless spacetime, and that it is also possible, with some reflection, to build a mental picture of that geometrically amorphous state.

Imagine the entities which are about to become assembled into spacetime and later into elements and elephants, as being a structureless dust. Now, at the time of when we speak, there is no spacetime, only the dust from which spacetime is to be built. The absence of spacetime, the absence of geometry, merely means that this point cannot be said to lie near or far from that; nor can this be said to precede this or follow that. Now there is absolute amorphousness. Later we shall have to sweep away the dust; but that will take care of itself, like all simplicities.

Before going any further, there is a point I want to make about changes like freezing, boiling, and the evaporation of frost and their relation to the patterns on wallpapers. The sharpness of such changes depends on the dimensionality of the space where they take place. In a one-dimensional universe, a drop of water would not freeze sharply, but would slowly solidify like butter as the temperature is lowered. In two dimensions the freezing would be sharper, and in three dimensions is sharper still. The sharpness of

The 7 frieze patterns, the 17 wallpaper patterns and the 230 space groups have all been described[90, 91, 92] in three as well as in higher dimensional spaces.[93]

this type of change can be investigated in spaces of different numbers of dimensions (even including spaces of fractions of dimensions). The sharpening of the changes can be traced as the number of dimensions increases smoothly through one, one and a half, two, and three, right up to four dimensions and beyond. At four dimensions the sharpness attains a peak and thereafter remains virtually the same.

The sharpness of changes like the ones we are considering arises from the collaboration of neighbours. When a molecule has very many neighbours their reorganization into a new type of material, such as the change from a liquid to a solid on freezing, is cooperative and proceeds swiftly. As the number of dimensions increases, the number of objects in the immediate neighbourhood of a point also increases. As a result the transitions become sharper. At four dimensions, though, each point has so many neighbours that any further increase is insignificant. With increasing dimensionality has gone increasing complexity, and that complexity is effectively complete when the dimensionality has reached four.

Some idea of the rapidity with which complexity increases with dimensionality can be obtained by thinking about patterns. A one-dimensional space is like a frieze, and it turns out that there are only seven basic types of pattern: all the friezes that have ever been made can be classified as one of seven types. A two-dimensional pattern is like a wallpaper. There are seventeen distinct patterns of wallpaper. Of course, there are very many more *designs*, because the underlying symmetry of the seventeen patterns can be displayed in various ways, with flowers of various kinds, peacocks, shapes, and colours. But all the great diversity of designs can be put into one of the seventeen classes. The increased complexity of two-dimensional space is reflected in the change from seven to seventeen. What of three-dimensional space? There are 230 different three-dimensional patterns. Another way of expressing this is that there are 230 basic types of crystal; 7, 17, 230, ... The number of patterns in four dimensions is about 5000. You can see how explosively this series is increasing, and that at four dimensions there is a greatly increased complexity of possible patterns.

An early description of the creation as a phase transition has been given[94] that has a number of features in common with the inflationary model.[57,95,96] The main thrust of the article concerns how the universe could have begun with equal abundances of matter and antimatter, yet have grown asymmetrical so that now the universe is composed wholly of matter. This is expressed as follows. 'One can speculate that the universe began in the most symmetrical state possible and that in such a state no matter existed: the universe was a vacuum. A second state was available, and in it matter existed. The second state had slightly less symmetry, but it was also lower in energy. Eventually a patch of the less symmetrical phase appeared and grew rapidly. The energy released by the transition found form in the creation of particles. This event might be identified with the "big bang".' For the development of this model into the inflationary universe, see the references cited above.

John Updike appears to have been captivated by this image, for it has motivated the closing pages of his novel *Roger's version*.[97]

The creation was like the freezing of water. At the creation the structureless dust of points grew into the ordering we now recognize as spacetime. Spacetime is four-dimensional because every point is then in a neighbourhood of sufficient complexity for it to have rich properties; properties corresponding to what we now recognize as particles and forces, and properties which have the consequence of conferring stability. Furthermore, the patterns—the knots—are trapped into persistence by the three dimensions of space, and are endowed with consequences for the future on account of the one dimension of time.

Spacetime emerged by chance out of its own dust. There was no need for intervention. Before time and place formed there were unrelated points, points that were not yet inter-related. They then lacked a geometry, and therefore were not yet spacetime.

Think of the primordial dust as swirling, and as swarming momentarily into clusters. Initially there are no relationships between the points: that point cannot be said to be next to that, or that one to precede that, for nearness and sequence do not yet have a meaning.

The swirling points may gather in sufficiently large numbers for relationships—patterns, knots—to be established, at least locally and temporarily. Sometimes a patch of this geometry may form outside time; sometimes there may be a patch of time fleetingly existing outside space. Sometimes patches of points patterned into regions of both space and time may come, and then scatter.

Chance is unlikely to constitute a patch of twelve-dimensional space and fifteen-dimensional time. Such a spacetime is unlikely to fluctuate into existence, because the degree of unassisted ordering required to establish such a complex structure is so great. But a one-dimensional strip of space or time (they are indistinguishable in one dimension) may flutter at hazard into existence. That existence arises through the haphazard, chance ordering of hitherto unrelated points, as real dust might blow into a line in the air. And like real dust it falls apart, for with its low dimensionality there is a poverty of neighbours and a concomitant lack of richness in its properties. The fleeting, one-dimensional universe that

It is not entirely clear what role 'unlikely' plays in this discussion. One may certainly re-express it as 'low statistical weight' as in *Gravitation*;[11] but if we are truly outside time, then it would appear that any event occurs whatever its likelihood so long as it is not absolutely impossible. If that is so, then universes are created 'all the while' and the present collection of universes is infinite and expanding infinitely rapidly. While vast, this scale of speculation is, of course, not the limit.

Small-scale fluctuations in the geometry of spacetime are taking place all round us (and inside us).[77] All the while and everywhere, the geometry of spacetime is crumbling and reforming, but it is taking place on so small a scale that we fail to notice it. The scale of length where the fluctuations are significant is the *Planck length*, which is equal to 1.6×10^{-33} cm. This suggests, in Wheeler's phrase,[46, 77] the 'foamlike nature of space'.

This is the subject of future physics. The picture I am giving here is necessarily vague because it is a speculation about the form that the final solution of the problem of the

chance bred, falls apart, and its incipient structure falls back into structurelessness.

Elsewhere (but there is no where yet) and at other times (but there is no time anywhere) the dust of spacetime happens to form itself into tiny universes of one dimension. But as they glitter into existence, they fail to survive, and leave no trace. Vast numbers of such still-born universes form. They constitute a place or establish an epoch; but fail, scatter, and die without a history.

The flickering, fleeting emergence of an incipient universe can be visualized as an aimless, purposeless, stumbling of points into a pattern. Among those vast numbers of chance efflorescences there are one or two (or a comparably small vast number) less likely and more complex clusters that constitute a two-dimensional universe. Many in that less likely vastness happen to assemble in a way that defines a two-dimensional space without time, and so are surface without duration. Others assemble, once again at hazard, into true two-dimensional spacetime, with a line of space and a direction of time. But they still lack enough complexity for survival. They are no more likely to survive than a cluster of motes in a sunbeam that briefly and transitorily happen to form a sheet. Such universes form, and outside our own space and time are forming now. Yet they are not destined to persist. As they come, so they go. They collapse into the structureless dust that chance happened to mould them from. They leave no trace in space or in time, for they formed their own space, and defined their own time, and when they die their spacetime dies with them.

One-dimensional universes were improbable patterns of points. Two-dimensional universes were yet more improbable patterns of the same points into more complicated, richer yet insufficiently rich, relationships. Much less probable still is the chance clustering that leads to spacetime of three dimensions. But still this spacetime is structurally too flimsy. The points are richer in neighbours than either of their two more probable predecessor universes (which, being outside space and time, after they have evaporated are not truly predecessors, and are forming and evaporating in another elsewhere and another now). But being not rich enough, they cannot survive their own formation. They fluctuate out of dust,

creation will take. Whereas it is possible to be precise and, with luck, even lucid about an established concept (because there is something to comprehend and then to convey) it is possible only to be vague about the events preceding the creation because they have not been established quantitatively. Nevertheless there are reasons why it would be unkind to regard these remarks as ludicrous and outside science. In the first place there must be some mechanism for the creation and its coming about. What we are attempting to express by these remarks is that there is the *possibility* of accounting for the creation, and the events that preceded it. We shall only have achieved the goal, though when this possibility has been expressed quantitatively. When that has been done I would like to think that its verbalization will be along the lines sketched in these pages. In a sense, it is merely a hunch, but one consistent with the entire thrust of modern science. Of course, if cosmogenesis does turn out to be like I describe (which I would find very pleasing), then this little essay would deserve no more credit than any other mythical account: all credit in science goes properly only to those who claw out every step forward by appealing to publicly assessed experiment and deeply argued mathematics. All armchair speculation, even when it scores a bullseye, is relatively contemptible.

The idea has been suggested that the existence of spacetime and the nature of geometry can be regarded as physics emerging from the statistics of long propositions.[77] That is the thought underlying the remarks about 'relationships': the possibility of concatenating entities into self-consistent networks. The particles are then the permissible self-consistencies, the networks then being the knots of spacetime. Only in three dimensions of space and one of time can this logical network take on sufficient richness (and stability). In the appendix referred to, the authors admit to failure; yet that does not mean that the idea is wrong. It may be appropriate to return to a problem left dangling in Chapter 5, where we saw the hint of an analogy between cosmogenesis and the emergence of mathematics. Although a system of numbers can be generated from the empty set, we are left with the nagging suspicion that a logic has been imposed in order to achieve the emergence. Now it may be that the logic somehow comes into being with the numbers, in a bootstrap procedure like that sketched here, and that the only mathematics we have is the one with a sufficiently universal self-consistency.

remain no more structured than a cloud of dust, and fall back into dust. Vast numbers of such three-dimensional universes come, by chance, and have gone back, through structural poverty, to dust again.

Then (whatever that means) by chance a clustering of the points stumbled into a pattern of such complexity that it corresponds to four dimensions; but they were four dimensions of space, and lacked time. That is rich in the complexity of inter-relationships, but not sophisticated enough for survival. Like so much dust the chance cluster of dust fell apart into structureless dust.

There was no wait for another four-dimensional pattern, because there is no waiting outside time. One of those patterns was four-dimensional spacetime. We know that it did in fact occur at least once. We may also suspect that it is continuing to occur outside our space and time; but our particular flutter of dust is the one with consequences for us. That particular fluctuation was the stumbling of the points into the pattern we discern as three dimensions of space and one dimension of time. By chance.

The fluctuation of which we are the ramification was four-dimensional, and was therefore rich in complexity in every neighbourhood. Moreover, its geometry was a geometry of spacetime. As such it was capable of supporting and sustaining the complexity of relations we interpret as matter and forces. Suddenly, by chance, we have a universe which is a collection of viable relationships. The relationships are sufficiently subtle and complex for the fluctuation to attain stability. Instead of whisping away like all the other improbabilities, this extreme improbability is frozen into existence. This particular universe survives. It is a seed for the organization of the whole of spacetime. We have been stumbled on. The universe has begun. By chance.

But what are these points? Where do they come from? Were they made, or did they emerge? Is there still a need to make something? Is infinite laziness unattainable?

We now stand at the eye of the creation. But we need one more concept, a concept able to account for the emergence of things out of nothing. I think it is discernible and would like to try to express how it might be comprehended.

While not wishing to jump too transparently on every currently successful bandwagon, it is interesting to note that the separation of nothing into opposites could bear an analogy to the formation of strings:[25, 57] the separation is like the emergence of a line that has two ends from a dimensionless point.

I quote from the same Appendix mentioned above: 'It is difficult to imagine a simpler element with which the construction of physics might begin then the choice yes—no or true—false or open—closed circuit. The combination of such elements gives one in the context of circuits a switching circuit ... It is isomorphic to a proposition in the propositional calculus of mathematical logic ...'[77] As we developed at length, and with ever increasing obscurity in Chapter 5, this analogy may be the deepest underlying reason why the universe can be described mathematically: mathematics is a system of logic and, if this view of the ultimate nature of the universe is correct, then it should not be surprising that mathematics can be applied to its description. In a sense mathematics mimics the ultimate structure. A formula written on a page is an expression of a particular group of relationships embedded in the structure of spacetime.

The key to the concept lies in noticing the cancellation of opposites. If the cancellation is thought of as reversed, then opposites separate from nothing. The world can be pictured as being constructed in such a distillation. At the creation, in some sense nothing has to separate into exceedingly simple opposites. If the separation generates a sufficiently complex pattern, the opposites acquire stability, and then richly persist.

A mundane example of this behaviour is the existence of matter and antimatter. A particle and its anti-particle on collision collapse into essentially nothing, a blob of energy; a particle and its anti-particle can be generated out of essentially nothing. The present universe is seething with this kind of activity, with energy (coiled spacetime) generating particles and anti-particles, and these collapsing back to energy again.

The creation, we have stressed, cannot spontaneously bring forth fully made up such rich complexities as particles and their anti-particles, let alone elephants and theirs. The creation can generate only the most primitive structures, structures of such simplicity that they can drop out from absolutely nothing. We still have to find the most simple.

The most simple is epitomized (I do not see how to apply a stronger term) by the difference between a point and no point, or between one and minus one. At heart the basis of the universe must be as simple as the difference symbolized by 1 and -1, or by yes and no, or (more prosaically) by true and false. The fundamental building blocks of the whole of creation must have this simple binary form. Nothing simpler has properties. Only the difference symbolized by 1 and -1, by one and not one, or point and no point, is sufficiently simple to be creatable, but rich enough when sufficiently concatenated (as in mathematics and in logic) to lead to properties. At root the universe is a dust of binary forms. That is the dust of spacetime.

But what brings into being the distinction of opposites?

I want to make one last but most important preliminary point before attempting to bring everything together and painting an overall picture. This point concerns a special role of time: time distinguishes opposites.

Creating things

The description of antiparticles in terms of particles evolving backwards in time[98] was introduced by E. C. G. Stückelberg, and developed by R. P. Feynman. It has been suggested that the reason why every electron in the universe is identical is that there is only one, and that we perceive a cross-section of its track as it weaves backwards and forwards in time, and therefore think it many. This remark is ascribed to J. A. Wheeler in the course of a telephone conversation with R. P. Feynman (or vice versa). [99] Such brave speculations, irrespective of their truth, epitomize the scientific attitude, if not its method.

Of course, I don't mean that an elephant and an anti-elephant can suddenly pop out of nothing. That is why I have stressed the important of the search for simplicities. Entities of *extreme* simplicity may be sufficiently simple to pop out of nothing in a manner we have yet to discuss.

The composition of discrete, two-component entities into a spacetime, the emergence of a geometry, is the basis of *twistor theory*.[100, 101]

We can work towards this concept by considering once again matter and antimatter. A particle and its antiparticle are distinguished by their direction of propagation in time. Instead of thinking of particles and antiparticles as separate, but suspiciously closely related species, we can think of an antiparticle as its particle counterpart travelling backwards in time. An electron travels forward in time; an anti-electron is an electron travelling backwards in time.

Opposites, at least of simple species, are distinguished by their direction of travel in time. This means that as soon as we have time we have the possibility of distinguishing the opposites which, when there is no time, blend into nothing.

Now we stand at the centre of the eye itself. Perhaps beyond the edge, but possibly just this side of the edge of understanding. I think we can see the rudiments of the self-inception of the world.

There are two ingredients. First we need the points that are to assemble into patterns defining space and time. Then we need the points that separate from their opposites by virtue of the pattern of time. Time lends life to the points; the points lend life to time. Time brought the points into being, and the points brought time into being. This is the cosmic bootstrap.

I should like to try to express this elusive idea (at least, *I* find it elusive; but just within mental grasp) in a slightly different way. We have represented opposites by the sumbols 1 and −1. As already suggested they can be taken as representing a point and its absence, or any sufficiently simple pair of opposites. 1 and −1, in the sense of standing for opposites, are distinguished by the direction they travel in time: − 1 is 1 travelling backwards in time. In the absence of time −1 and 1 coalesce into—well, into $1 - 1 = 0$, or nothing.

The elusive, yet just graspable central speculation is that the universe comes into existence by virtue of self-reference. We have argued that 1 and −1, the points and their absence, constitute time and space when appropriately arrayed. But in order to exist and to come into being, point and no point need time, for time separates them, distinguishes them, and induces them from nothing. There

is the central self-reference: the emergence of time from its dust; dust brought into being by the act of patterning time.

In a word, the central speculation is that spacetime generates its own dust in the process of its own self-assembly. The universe can emerge out of nothing, without intervention. By chance.

Sixth Orientation

I have argued that there is no need to regard ourselves as anything other than the ramifications of chance. The universe could have stumbled upon its own existence as other universes perhaps continue to stumble outside our space and time, defining their own space and their own time. I have tried to show how the flutter into a pattern that possessed enough complexity for stability was the stumbling that need have involved no intervention. It brought us about (and in due course will see to our disinheritance). We can even begin to discern how the universe could come from absolute nothing as time induced (by chance) its own existence.

That is really the end of our journey. We have been back to the time before time, and have tracked the infinitely lazy creator to his lair (he is, of course, not there). We have seen, even if without sharp clarity, the coming into being of things from nothing. In a broad sense we have accounted for the nature of being in terms of the nature of spacetime and the interdependent consequences of its unwinding as it collapses without purpose into chaos. We are in the midst of this unwinding.

We can now collect these speculations together and sweep in imagination through this half of eternity. We shall begin before the beginning, allow unbounded speculation, and pursue the flight of the universe beyond its future.

Seven

CREATED THINGS

FIRST, THERE IS THE BEGINNING.

In the beginning there was nothing. Absolute void, not merely empty space. There was no space; nor was there time, for this was before time. The universe was without form and void.

By chance there was a fluctuation, and a set of points, emerging from nothing and taking their existence from the pattern they formed, defined a time. The chance formation of a pattern resulted in the emergence of time from coalesced opposites, its emergence from nothing. From absolute nothing, absolutely without intervention, there came into being rudimentary existence. The emergence of the dust of points and their chance organization into time was the haphazard, unmotivated action that brought them into being. Extreme simplicities, emerged from nothing.

Yet the line of time collapsed, and the incipient universe evaporated, for time alone is not rich enough for existence. Time and space emerged elsewhere, but they too crumbled back into their own dust, the coalescence of opposites, or simply nothing.

Patterns emerged again, and again, and again. Each time the pattern formed a time, and through their patterning into time, the points induced their own existence. Sometimes chance led to a pattern forming two dimensions of what we would recognize as time; but because backwards was then attainable from forwards, opposites were not distinct. There was no stability, and the opposites blended back into nothing.

Sometimes chance patterned points into a space as well as a time; but there was not room for complexity, and the pattern chance stumbled upon crumbled. It lost time, and in losing time it lost its existence.

Then, by chance, there came about our fluctuation. Points came into existence by constituting time but, this time, in this pattern time was accompanied by three dimensions of space. A geometry was born that has complexity and sophistication. Its complexity arises from the density of its neighbours and gives it a sophistication sufficient for the existence of matter, energy, and forces; and

with them comes stability, later elements, and still later elephants. This fluctuation, we notice, survives.

The initial generation of spacetime left it coiled and twisted. The local twists into persistent knots are the particles that now constitute the things like elephants. Different species of particle are different species of knots in the structure of spacetime. Different knots are different groupings (like everyday knots are different ravellings of string) of the binary entities coming into existence at the creation. Different particles are therefore different local topological structures of spacetime. Embedding these local structures in spacetime has distant consequences. In particular it induces the phenomenon of gravitation, the global twist of spacetime.

The tendency of the universe is to attain global uniformity, a three-dimensional flatness. Energy, which includes matter, is coiled spacetime. Coiled spacetime is the spring of the universe, and our activities, like all activity, are aspects of its unwinding. The evolution of the universe is the dispersal of the ripples in spacetime.

There can be two terminations of the future.

In one, the knots loosen and untie, their local ripples dissipate, and spacetime in due course becomes everywhere uniformly and eternally flat. The universe continues to exist, but lies unwound. It lacks activity, and is uniformly and irretrievably dead. All traces of our achievement—our art and our knowledge—will have been obliterated and will be as though they never were.

In another, there may be so many local knots, and their untying so low, that their combined and total distant twisting winds back the universe to a point, perhaps to bounce out again. That bounce is not creation, but renewal. Indeed, we may be living even now in a renewed universe, with the true creation generations of universes ago. Periodically reconstituted universes may constitute a perpetuity for the future, but in the ultimate past they must once have sprung from a true creation (unless time lies in a circle).

And finally there is the present.

Currently our universe is alive. Its life—its activity in all its forms—is permitted by the balance of the strengths of the forces that govern motion, that constitute atoms, and that bind atoms into elephants and galaxies.

Quark and antiquark pairs exist, and transmit the strong force. But only trios of quarks (or of antiquarks) have been observed. The existence of quarks is no longer seriously in doubt: experiments on protons have shown that they have an inner structure which behaves as three quarks should. In particular, the quarks appear to act like dimensionless points (like the electron); [10, 23, 25, 57] there are also speculations on inner structure.[57, 102]

There has been virtually no progress with the deduction of the values of the fundamental constants. That might mean that there is no progress to be made, as there is no progress to be made with accounting for the value of the factor π for converting a radius into a circumference (but showing that they were intrinsically topological parameters would, of course, be progress). Many people have come up with combinations of π that gives almost the right experimental value of α, but then there are statistically numerous ways of constructing numbers close to 1/137.0360. The simplest expression for α, which also gives the 1992 experimental value (to seven significant figures), is $1/\sqrt{(137^2 + \pi^2)}$, but that might also be a coincidence. The problem may be akin to 'explaining' the value of the golden ratio,[103] $\varphi = \frac{1}{2}(1 + \sqrt{5}) = 1.61803$.

There are the deepest forces of all, the forces that bind together the most basic components of the discernible world, the quarks. Quarks, or at least their conjectured components, seem to be devoid of any deeper structure, which is why we can begin to believe that the onion now lies unpeeled. They are existence without extent. They invariably occur in threes, and the force between them binds them so securely that separating them seems a prospect as remote as separating the three dimensions of space: that may indeed be the reason for their inseparability. At the quarks we are at, or almost at, the most primitive manifestation of the underlying concatenation of fundamental forms, and disentangling three quarks may be as impossible as disentangling space.

There are also other forces. There is gravitation, which acts between everything—between lumps of energy and matter as well as between different lumps of matter—and which feebly yet unavoidably and ubiquitously binds the universe into an entity. There is also the electric force, which weakly binds electrons to nuclei and forms atoms, the delicate structures having the responsiveness and malleability suitable for the evolution of the subtle property of life. There are also the strong forces and the weak forces, the forces that operate between the elementary particles, the fundamental knots of spacetime.

The balance of the strengths of these forces is crucial to the emergence of conscious life, although alien and unknowing universes may litter the void and have just as little purpose as ours.

If nuclei were bound together slightly more weakly, or slightly more strongly, the universe would lack a chemistry; and life, apparently biology but truly physics in the form of chemistry, would be absent. If the electric force were slightly stronger than it is, evolution would not reach organisms before the sun went out. If it were only slightly less, stars would not have planets, and life would be unknown.

That such a universe as ours did emerge with exactly the right blend of forces may have the flavour of a miracle, and therefore seem to require some form of intervention. But nothing intrinsically lacks an explanation. We cannot yet see quite far enough to decide which is the right explanation, but we can be confident that

As mentioned above, whether or not the universe is open or closed is itself an open question, although the view is shifting towards it being flat. That would fit better with the view that a finite amount of matter was generated at the creation. An open, eternal, universe would also be of infinite spatial extent at all times, even at the beginning, if there was a beginning.[104]

The role of the magnitudes of the fundamental constants, especially of the extent to which they permit the evolution of consciousness to the point of being able to wonder about them, has been investigated.[21, 22, 105] So too has the possibility that they change with time.[106, 107] Carter,[105] for example, makes the point that the division of the main sequence stars into blue giants and red dwarfs depends on a critical coincidence between the strengths of the electromagnetic and gravitational interactions. If gravity were only a little stronger (or α only marginally weaker) all the main sequence stars would be blue giants. Planetary formation appears to involve red dwarfs, and so planets, and presumably consciousness, would not have emerged. Carr and Rees[21] treat the matter in more detail, and deal with the sizes of planets, mountains, and men and their injuries, as well as the distribution of the elements. The entire field of modern anthropocentricity is wonderfully and thoroughly surveyed by Barrow and Tipler.[22]

intervention was not necessary. Chance may have resulted in a benevolent job lot of strengths of forces. If they had not been so benevolent the universe would have gone along very well all the same, but may have lacked stars and planets, or have passed in a moment, or have been eternally dense. We would then have been none the wiser, and none the sadder, because we would not have been. But could chance acting alone have been so fortunate?

Chance might have stumbled on fortune. Not at first but in due course, for a universe looping through cycles of existence might rebound each time with a different bag of strengths of forces. In our cycle of the universe, which may be the first, but could be the ten-thousand million millionth or so since the true creation, the recycled structure of spacetime supports a web of forces that happens to be exactly right for the emergence of consciousness. This universe was constituted, or reconstituted, by chance to come awake, as it may have been countless times before and may be again in the future. There may have been previous universes lacking self-awareness, and others even more devoid of feature. Happily they collapsed and we have our turn, as other's turns may yet come.

The universe might be a single shot. One creation, one coiling of the spring, and one irreversible drift into uncoiled uniformity and global flatness. Perfect ultimate flatness, lacking activity and expectation of revived activity. Dead flat spacetime.

In such a universe there is still no purpose behind the benevolence of the forces. It might be chance that has given them, the forces, their strengths and we are the beneficiaries, not knowing otherwise if things had been otherwise, alive through chance. It may be that the strengths of the forces change with time, and that we live in a universe during an epoch when they happen to be kind. The universe has come awake during this the epoch of benevolence; consciousness has emerged, not because it was needed but simply because it happened, and the universe will return to its sleep when the epoch has passed and the forces have taken on new strengths. We, we the universe, are awake only now, and necessarily we are awake amid benevolence.

It is possibly the case that the generation of spacetime in a

Fundamental science may be almost at an end, and might be completed within a generation. Such views have been held before, but people then mistook smallness for simplicity. Only when structure has been peeled back to the point where it ceases to need further structure—when it has been peeled back to the point of extreme simplicity, such as lack of spatial extent—can we be confident that we have reached the end. Only when everything can be explained in exceedingly simple terms, when everything almost literally falls into place without us having to account for its properties, will fundamental science be on the point of resting.

That is not to say that science as a whole need ever sleep. There are extremely difficult and important questions, such as those concerning the details of biological function, that will remain, at least for a few hundred years. Such questions involve explorations of the branches of the tree of knowledge, but the fundamental nature of the world, the roots of the tree, will soon be qualitatively and quantitatively certain.

transition from the void necessarily leads to the strengths of forces as we know them, for forces are aspects of the structure of spacetime. That still does not imply a purpose; we can still remain the children of aimless chance. The forces, the fundamental constants of nature like the speed of light and the strength of the electric charge, may be no more significant than the structure, or our description of the structure, of spacetime, and we should no more wonder at their values than we wonder at the value of 1.609 344 km per mile or at the value of π. To be sure that this is the explanation we need to turn to the next page of physics.

When we have dealt with the values of the fundamental constants by seeing that they are unavoidably so, and have dismissed them as irrelevant, we shall have arrived at complete understanding. Fundamental science then can rest. We are almost there. Complete knowledge is just within our grasp. Comprehension is moving across the face of the Earth, like the sunrise.

BIBLIOGRAPHY

1. 'Evolution', *Scientific American* **239**, No. 3 (September 1978)
2. R. Dawkins, *The selfish gene*, 2nd edn, Oxford University Press (1989)
3. R. Shapiro, *Origins: a skeptic's guide to the creation of life on earth*, Heinemann (1986)
4. S. F. Mason, *Chemical evolution*, Oxford University Press (1991)
5. L. Stryer, *Biochemistry*, 3rd edn, W. H. Freeman (1988). The first eight chapters of this book are available as L. Stryer, *Molecular design of life*, W. H. Freeman (1989)
6. P. W. Atkins, *Atoms, electrons, and change*, Scientific American Library (1991)
7. S. Mitton (ed.), *The Cambridge encyclopaedia of astronomy*, Jonathan Cape (1977)
8. P. A. Cox, *The elements: their origin, abundance, and distribution*, Oxford University Press (1989)
9. W. J. Kaufmann, *Universe*, W. H. Freeman, New York (1991)
10. N. Calder, *The key to the universe*, BBC Publications (1977).
11. C. W. Misner, K. S. Thorne and J. A. Wheeler, *Gravitation*, W. H. Freeman (1973)
12. M. Rowan-Robinson, *Cosmology*, 2nd edn, Oxford University Press (1981)
13. P. J. E. Peebles, *Physical cosmology*, Princeton University Press (1971)
14. S. L. Jaki, 'Olbers', Halley's, or whose paradox?', *American Journal of Physics* **35**, 200–210 (1961)
15. P. T. Landsberg and D. A Evans, *Mathematical cosmology*, Clarendon Press (1977)
16. E. Harrison, 'Olbers paradox', *Nature* **352**, 574 (1991)
17. P. W. Atkins, *Physical chemistry*, 4th edn, Oxford University Press and W. H. Freeman (1990)
18. P. W. Atkins, *Quanta: a handbook of concepts*, 2nd edn, Oxford University Press (1991)
19. E. Schrödinger, *What is life?*, Cambridge University Press (1969)
20. F. J. Dyson, 'Time without end: physics and biology in an open universe', *Reviews of Modern Physics* **51**, 447–460 (1979)
21. B. J. Carr and M. J. Rees, 'The anthropic principle and the structure of the physical world', *Nature* **278**, 605–612 (1979)
22. J. D. Barrow and F. J. Tipler, *The anthropic cosmological principle*, Oxford University Press (1986)

23. J. C. Polkinghorne, *The particle play*, W. H. Freeman (1979)
24. A. Pais, *Inward bound: of matter and forces in the physical world*, Oxford University Press (1986)
25. P. D. B. Collins, A. D. Martin and E. J. Squires. *Particle physics and cosmology*, Wiley (1989).
26. P. W. Atkins, *The second law*, Scientific American Library (1984)
27. F J. Dyson, 'Energy in the universe', *Scientific American* **225**, No. 3, 50–59 (1971)
28. B. B. Mandelbrot, *The fractal geometry of nature*, W. H. Freeman (1982)
29. G. Nicolis and I. Prigogine, *Exploring complexity: an introduction*, W. H. Freeman (1989)
30. P. C. W. Davies, *The physics of time asymmetry*, Surrey University Press (1974)
31. J. D. Watson, *The double helix: a personal account of the discovery of the structure of DNA*; critical annotated version, G. Stent. Weidenfeld and Nicolson (1981)
32. J. E. Darnell, D. Baltimore and H. Lodish, *The molecular biology of the cell*, Scientific American Books (1990)
33. J. P. Changeux, *Neuronal man: the biology of mind*, Oxford University Press (1985)
34. J. Z. Young, *Programs of the brain*, Oxford University Press (1978)
35. S. Rose, *The conscious brain*, Penguin books (1976); Vintage books (1976)
36. S. Snyder, *Drugs and the brain*, Scientific American Library (1986)
37. E. R. Kandel, 'Small systems of neurons', *Scientific American* **241**, No. 3, 60–70 (1979).
38. W. Yourgrau and S. Mandelstam, *Varation principles in dynamics and quantum theory*, Pitman (1968)
39. 'Light', *Scientific American* **219**, No. 3 (September 1968).
40. K. Naussau, *The physics and chemistry of color*, Wiley (1983)
41. R. P. Feynman and A. R. Hibbs, *Quantum mechanics and phase integrals*, McGraw-Hill (1965).
42. R. P. Feynman, R. B. Leighton and M. Sands, *The Feynman lectures in physics*, Vol. 3, Addison-Wesley (1963)
43. W. J. Kaufmann III, *Black holes and warped spacetime*, W. H. Freeman (1979)
44. J. C. Graves, *The conceptual foundations of contemporary relativity theory*, MIT Press (1971)
45. J. Schwinger, *Einstein's legacy*, Scientific American Library (1986)
46. J. A. Wheeler, *Gravity and spacetime*, Scientific American Library (1990)
47. L. P. Hughston and K. P. Tod, *An introduction to general relativity*, Cambridge University Press (1990)
48. G. J. Whitrow, *The natural philosophy of time*, 2nd edn, Clarendon Press (1980)
49. K. G. Denbigh, *Three concepts of time*, Springer-Verlag (1981)
50. P. V. Coveney and R. Highfield, *The arrow of time*, W. H. Allen (1990)
51. R. Flood and M. Lockwood, *The nature of time*, Basil Blackwell (1986)

52. E. F. Taylor and J. A. Wheeler, *Spacetime physics*, 2nd edn, W. H. Freeman (1992)

53. H. P. Robertson, 'Postulate versus observation in the special theory of relativity', *Reviews of Modern Physics* **21**, 378–382 (1949)

54. W. L. Burke, *Spacetime, geometry, cosmology*, University Science Books (1980)

55. I. Prigogine, *From being to becoming*, W. H. Freeman (1980)

56. P. C. W. Davies, *The forces of nature*, Cambridge University Press (1979)

57. P. C. W. Davies, *The new physics*, Cambridge University Press (1989)

58. D. Z. Freedman and P. Van Nieuwenhuizen, 'Supergravity and the unification of the laws of physics', *Scientific American* **238**, No. 2, 126–143 (1978)

59. D. R. Hofstadter, *Gödel, Escher, Bach: an Eternal Golden Braid*, Basic (1979); Penguin (1980)

60. S. E. Kim, 'The impossible skew quadrilaterial: a four dimensional optical illusion' in *Proceedings of the American Association for the Advancement of Science Symposium on Hypergraphics: Visualizing complex relationships in art and science* (ed. D. Brisson), Westview Press (1978)

61. M. Gardner, 'Mathematical games', *Scientific American* **243**, No. 1, 14–20 (1980), and references therein.

62. A. K. Dewdney, *Two-dimensional science and technology*, Pre-print; Dept. of Computer Science, University of Western Ontario (1980)

63. A. K. Dewdney, *The planiverse*, Pan Books (1984).

64. G. J. Whitrow, 'Why physical space has three dimensions', *British Journal of the Philosophy of Science* **6**, 13–31 (1955)

65. G. L. Whitrow, *The structure and evolution of the universe*, Harper (1959)

66. T. L. Saty, 'Operations analysis' in *The mathematics of physics and chemistry* (eds H. Margenau and G. M. Murphy), **2**, 249–320, Van Nostrand (1964)

67. P. M. Morse and H. Feshbach, *Methods of theoretical physics*, Vo. 1, McGraw-Hill (1953)

68. P. Ehrenfest, 'In what way does it become manifest in the fundamental laws of physics that space has three dimensions?', *Proceedings of the Amsterdam Academy* **20**, 200–209 (1917)

69. P. C. W. Davies, *Other worlds*, Dent (1980)

70. L. Neuwith, 'The theory of knots', *Scientific American* **240**, No. 6, 84–96 (1979)

71. R. Courant and H. Robbins, *What is mathematics?*, Oxford University Press (1941)

72. C. Rebbi, 'Solitons', *Scientific American* **240**, No. 2, 76–91 (1979)

73. Z. Parsa, 'Topoligical solitons in physics', *American Journal of Physics* **47**, 56–62 (1979)

74. P. Collas, 'General relativity in two-and three-dimensional space-times', *American Journal of Physics* **45**, 833–837 (1977)

75. R. Penney, 'On the dimensionality of the real world', *Journal of Mathematical Physics* **6**, 1607–1611 (1965)

76. J. Dorling, 'The dimensionality of time', *American Journal of Physics* **38**, 539–540 (1970)

77. C. M. Patton and J. A. Wheeler, 'Is physics legislated by cosmogony?' in *Quantum gravity* (eds C. J. Isham, R. Penrose and D. W. Sciama), 538–605, Clarendon Press, 1975; and also in *The encyclopaedia of ignorance* eds R. Duncan and M. Weston Smith), 19–35, Pergamon (1977).

78. R. W. Livingstone, ed., *The legacy of Greece*, Oxford University Press (1921)

79. A. Pais, *Subtle is the Lord*, Oxford University Press (1982)

80. J. A. Wheeler and W. H. Zurek, eds., *Quantum theory and measurement*, Princeton University Press (1983)

81. J. D. Barrow, *Theories of everything*, Oxford University Press (1990); Vintage (1992)

82. J. Lyons, *Chomsky*, Fontana Modern Masters (1970)

83. J. H. Conway, *On numbers and games*, Academic Press (1976)

84. H. O. Peitgen and D. Saupe, eds., *The science of fractal images*, Springer-Verlag (1988)

85 .H. O. Peitgen and P. H. Richter, *The beauty of fractals*, Springer-Verlag (1986)

86. M. Schroeder, *Fractals, chaos, and power laws*. W. H. Freeman (1991)

87. P. J. E. Peebles, D. N. Schramm, E. L. Turner and R. G. Kron, 'The case for the relativistic hot Big Bang cosmology', *Nature* **352**, 769 (1991)

88. L. M. Lederman and D. R. Schramm, *From quarks to the cosmos*, Scientific American Library (1989)

89. K. G. Wilson, 'Problems in physics with many scales of length', *Scientific American* **241**, No. 2, 140–157 (1979)

90. H. Weyl, Symmetry, Princeton University Press (1952)

91. J. Rosen, *Symmetry discovered*, Cambridge University Press (1975)

92. E. H. Lockwood and R. H. Macmillan, *Geometric symmetry*, (1978)

93. R. L. E. Schwarzenberger, *N-dimensional crystallography*, Pitman (1980)

94. F. Wilczek, 'The cosmic asymmetry between matter and antimatter', *Scientific American* **243**, No. 6, 60–68 (1980)

95. A. H. Guth and P. J. Steinhardt, 'The inflationary universe', *Scientific American* 90–102 (May 1984)

96. E. P. Tryon, 'Cosmic inflation', in *The encyclopedia of physical science and technology*, Vol. 3 709. Academic Press (1987)

97. J. Updike, *Roger's version*, Deutsch (1986)

98. M. Gardner, 'Can time go backward?', *Scientific American* **216**, No. 1, 98–108 (1967)

99. J. R. Lucas, *A treatise on time and space*, Methuen, p. 44 (1973)

100. R. Penrose, 'Angular momentum: an approach to combinatorial spacetime' in *Quantum theory and beyond* (ed. T. Bastin), 151–180, Cambridge University Press (1971)

101. R. Penrose, *The emperor's new mind*, Oxford University Press (1990)

102. See *Scientific American* **244**, No. 2, 64–68 (1981)

103. H. E. Huntley, *The divine proportion*, Dover (1970)

104. S. W. Hawking, *A brief history of time: from the Big Bang to black holes*, Bantam (1988)

105. B. Carter, *Confrontation of cosmological theories with observational data* (ed. M. S. Longair), Reidel (1974)

106. F. J. Dyson, 'The fundamental constants and their times variation' in *Aspects of quantum theory* (eds A. Salam and E. P. Wigner), 213–236, Cambridge University Press (1972)

107. P. S. Wesson, *Cosmology and geophysics*, Hilger (1978)

READ MORE IN PENGUIN

In every corner of the world, on every subject under the sun, Penguin represents quality and variety – the very best in publishing today.

For complete information about books available from Penguin – including Puffins, Penguin Classics and Arkana – and how to order them, write to us at the appropriate address below. Please note that for copyright reasons the selection of books varies from country to country.

In the United Kingdom: Please write to *Dept. JC, Penguin Books Ltd, FREEPOST, West Drayton, Middlesex UB7 OBR*

If you have any difficulty in obtaining a title, please send your order with the correct money, plus ten per cent for postage and packaging, to *PO Box No. 11, West Drayton, Middlesex UB7 OBR*

In the United States: Please write to *Penguin USA Inc., 375 Hudson Street, New York, NY 10014*

In Canada: Please write to *Penguin Books Canada Ltd, 10 Alcorn Avenue, Suite 300, Toronto, Ontario M4V 3B2*

In Australia: Please write to *Penguin Books Australia Ltd, 487 Maroondah Highway, Ringwood, Victoria 3134*

In New Zealand: Please write to *Penguin Books (NZ) Ltd, 182–190 Wairau Road, Private Bag, Takapuna, Auckland 9*

In India: Please write to *Penguin Books India Pvt Ltd, 706 Eros Apartments, 56 Nehru Place, New Delhi 110 019*

In the Netherlands: Please write to *Penguin Books Netherlands B.V., Keizersgracht 231 NL–1016 DV Amsterdam*

In Germany: Please write to *Penguin Books Deutschland GmbH, Friedrichstrasse 10–12, W–6000 Frankfurt/Main 1*

In Spain: Please write to *Penguin Books S. A., C. San Bernardo 117–6° E–28015 Madrid*

In Italy: Please write to *Penguin Italia s.r.l., Via Felice Casati 20, I–20124 Milano*

In France: Please write to *Penguin France S. A., 17 rue Lejeune, F–31000 Toulouse*

In Japan: Please write to *Penguin Books Japan, Ishikiribashi Building, 2–5–4, Suido, Tokyo 112*

In Greece: Please write to *Penguin Hellas Ltd, Dimocritou 3, GR–106 71 Athens*

In South Africa: Please write to *Longman Penguin Southern Africa (Pty) Ltd, Private Bag X08, Bertsham 2013*

READ MORE IN PENGUIN

SCIENCE AND MATHEMATICS

QED   Richard Feynman
The Strange Theory of Light and Matter

'Physics Nobelist Feynman simply cannot help being original. In this quirky, fascinating book, he explains to laymen the quantum theory of light – a theory to which he made decisive contributions' – *New Yorker*

Does God Play Dice?   Ian Stewart
The New Mathematics of Chaos

To cope with the truth of a chaotic world, pioneering mathematicians have developed chaos theory. *Does God Play Dice?* makes accessible the basic principles and many practical applications of one of the most extraordinary – and mind-bending – breakthroughs in recent years.

Bully for Brontosaurus   Stephen Jay Gould

'He fossicks through history, here and there picking up a bone, an imprint, a fossil dropping and, from these, tries to reconstruct the past afresh in all its messy ambiguity. It's the droppings that provide the freshness: he's as likely to quote from Mark Twain or Joe DiMaggio as from Lamarck or Lavoisier' – *Guardian*

The Blind Watchmaker   Richard Dawkins

'An enchantingly witty and persuasive neo-Darwinist attack on the anti-evolutionists, pleasurably intelligible to the scientifically illiterate' – Hermione Lee in the *Observer* Books of the Year

The Making of the Atomic Bomb   Richard Rhodes

'Rhodes handles his rich trove of material with the skill of a master novelist ... his portraits of the leading figures are three-dimensional and penetrating ... the sheer momentum of the narrative is breathtaking ... a book to read and to read again' – Walter C. Patterson in the *Guardian*

Asimov's New Guide to Science   Isaac Asimov

A classic work brought up to date – far and away the best one-volume survey of all the physical and biological sciences.